AF759300

ELECTRICS VERSUS HYDRAULICS VERSUS PNEUMATICS

200039549

Conference Planning Panel

R Walters, BSc
Consultant Engineer
Wembley
Middlesex

Professor D E Bowns, BSc(Eng), PhD, CEng, MIMechE, FInstMC
School of Engineering
University of Bath
Bath
Avon

R M Wells, BSc
Association of Hydraulic Equipment Manufacturers
London

G Allison, BSc, MSc, CEng, MIMechE
Pilkington Brothers
Lathom
Lancashire

J R Archer, AIMechE
P A Technology
Melbourn
Royston
Hertfordshire

G Le Hunte, CEng, MIMechE, MIProdE
Compair Maxam Limited
Redruth
Pool
Cornwall

G B Smith, BSc, CEng, MIEE
British Railways Board
The Railway Technical Centre
Derby
Derbyshire

EUROPEAN CONFERENCE ON

ELECTRICS VERSUS HYDRAULICS VERSUS PNEUMATICS

IMechE CONFERENCE PUBLICATIONS 1985–11

Sponsored by
The Engineering Manufacturing Industries Division of
The Institution of Mechanical Engineers
Association of Hydraulic Equipment Manufacturers
British Compressed Air Society
British Hydromechanics Research Association
European Oil Hydraulic and Pneumatic Committee (CETOP)
The Institution of Electrical Engineers

22 October 1985
1 Birdcage Walk
Westminster
London SW1

MEP

Published for
The Institution of Mechanical Engineers
by Mechanical Engineering Publications Limited
LONDON

First published 1985

This publication is copyright under the Berne Convention and the International Copyright Convention. Apart from any fair dealing for the purpose of private study, research, criticism or review, as permitted under the Copyright Act 1956, no part may be reproduced, stored in a retrieval system, or transmitted in any form or by any means, electronic, electrical, chemical, mechanical, photocopying, recording or otherwise, without the prior permission of the copyright owners. Inquiries should be addressed to: The Managing Editor, Mechanical Engineering Publications Limited, PO Box 24, Northgate Avenue, Bury St Edmunds, Suffolk IP32 6BW

© The Institution of Mechanical Engineers 1985

ISBN 0 85298 568 1

200039549
SHEFFIELD UNIV.
APPLIED SCIENCE
LIBRARY
16.XII.85

The Publishers are not responsible for any statement made in this publication. Data, discussion and conclusions developed by authors are for information only and are not intended for use without independent substantiating investigation on the part of potential users.

Printed by Waveney Print Services Ltd, Beccles, Suffolk

CONTENTS

The Institution of Mechanical Engineers

The primary purpose of the 76,000-member Institution of Mechanical Engineers, formed in 1847, has always been and remains the promotion of standards of excellence in British mechanical engineering and a high level of professional development, competence and conduct among aspiring and practising members. Membership of IMechE is highly regarded by employers, both within the UK and overseas, who recognise that its carefully monitored academic training and responsibility standards are second to none. Indeed they offer incontrovertible evidence of a sound formation and continuing development in career progression.

In pursuit of its aim of attracting suitably qualified youngsters into the profession — in adequate numbers to meet the country's future needs — and of assisting established Chartered Mechanical Engineers to update their knowledge of technological developments — in areas such as CADCAM, robotics and FMS, for example — the IMechE offers a comprehensive range of services and activities. Among these, to name but a few, are symposia, courses, conferences, lectures, competitions, surveys, publications, awards and prizes. A Library containing 150,000 books and periodicals and an Information Service which uses a computer terminal linked to databases in Europe and the USA are among the facilities provided by the Institution.

If you wish to know more about the membership requirements or about the Institution's activities listed above — or have a friend or relative who might be interested — telephone or write to IMechE in the first instance and ask for a copy of our colour 'at a glance' leaflet. This provides fuller details and the contact points — both at the London HQ and IMechE's Bury St Edmunds office — for various aspects of the organisation's operation. Specifically it contains a tear-off slip through which more information on any of the membership grades (Student, Graduate, Associate Member, Member and Fellow) may be obtained.

Corporate members of the Institution are able to use the coveted letters 'CEng, MIMechE' or 'CEng, FIMechE' after their name, designations instantly recognised by, and highly acceptable to, employers in the field of engineering. There is no way other than by membership through which they can be obtained!

NEW ○ NEW ○ NEW ○ NEW ○ NEW

PNEUDRI™

Compressed Air Dryers For The 90's

- **HIGH QUALITY CLEAN, OIL FREE AND DRY COMPRESSED AIR**
- **ENERGY SAVING** system combined with simple operation.
- **COMPACT AND LIGHTWEIGHT**
- **EASY INSTALLATION WITH LOW MAINTENANCE**
- **HIGH EFFICIENCY INTEGRAL INLET/ OUTLET FILTRATION**
- **RELIABLE MICROPROCESSOR** control and monitoring.
- **FLEXIBLE** modular design facilitates further plant expansion.

domnick hunter filters limited

Durham Road,
Birtley, Co. Durham,
England DH3 2SF.

Tel: 091 410 5121
Telex: 537282
Telefax: 091 410 7621

Fork lift trucks – an ideal application for regenerative transmissions

D S MACARTHUR, AMI MechE
PA Technology, Melbourn, Royston, Hertfordshire

SYNOPSIS The counterbalance fork-lift truck is an ideal application for regeneration of braking energy. The combination of a high mass and an arduous duty gives high values of percentage energy recoverable (%ER). The %ER depends strongly on the duty but can easily be measured, and is typically 50 - 85%. Electrical regeneration is simple but ineffective for various reasons. Hydrostatic regeneration is shown to have much larger potential energy savings coupled with improved performance. Several possible configurations are outlined and discussed.

1 INTRODUCTION

Regeneration of the energy dissipated in braking a wheeled vehicle has long been the ambition of many energy-conscious engineers. It is frustrating to realise that every time you brake your car from even a moderate speed, you throw away energy worth several pence. Could not this energy be recovered and re-used?

Many research and development projects with this aim have been carried out over the last 20 years, studying electrical, flywheel and hydraulic systems, but have so far resulted in relatively little working hardware, and certainly no system viable for that voracious energy consumer, the private car. The reasons for this are common to many superficially attractive energy-saving ideas which fail to fulfil their initial promise:

- the energy savings achievable in practice are much smaller than predicted by simple theoretical models
- other performance aspects are generally compromised
- the capital cost is too high
- energy costs are still too low to justify the capital costs and performance compromise.

Fork-lift trucks however, have a very high mass, (typically 4 - 10 tonne), relatively low top speeds (typically 15 - 22 km/hr), and make very frequent stops and starts. As a result, a larger proportion of the propulsion energy is dissipated in braking than for other vehicles. The actual proportion varies with the duty cycle, the type of truck, and driving technique, but is typically over 50%, and may be as much as 85% of the propulsion energy.

Recovery of this wasted energy is an attractive proposition, and offers benefits other than the obvious savings in energy costs. In particular, with an electric fork lift truck, the real cost of electrical energy from a storage battery is very high, when battery depreciation and charging costs are included. Thus savings in propulsion energy demand may be realised in any of several ways:

- a smaller (cheaper) battery for the same duty;
- extended range with the same battery;
- shallower discharges and hence longer battery life;
- shorter charging times or reduced charging currents.

Other potential benefits, applicable to i.c. engined trucks as well as electric trucks, include:

- reduced brake wear and maintenance costs
- improved acceleration performance
- reduced engine/drive motor size and cost
- reduced fuel consumption
- with some concepts, increased lift speed.

Thus an effective regenerative transmission could offer improved performance as well as energy and other cost savings. As performance in a fork lift truck translates directly into productivity, the economic justification is good.

2 ENERGY CONSUMPTION OF A FORK LIFT TRUCK

The energy consumption can be divided into:

- propulsion energy
- lift energy
- auxiliary energy (e.g. power steering, mast tilt).

Propulsion energy is usually much greater than the other two, but the proportions depend

C275/85 © IMechE 1985

on the duty cycle, and it is quite possible for the lift energy to be comparable to the propulsion energy, particularly in high-lift, low-speed electric trucks.

Propulsion energy is required

- to accelerate the vehicle (i.e. increase its Kinetic Energy (K.E.))

- to propel it up gradients (i.e. increase its Potential Energy (P.E.))

- to overcome aerodynamic drag

- to overcome tyre rolling resistance.

The first two are conservative (i.e. in principle, the energy is available for recovery) while the latter two are dissipative.

For a fork-lift truck, aerodynamic drag is quite negligible. The effect of even quite small gradients can be very significant, but most industrial trucks operate on very level floors, and a simplified analysis of energy consumption can ignore gradients.

The tyre rolling resistance is a drag force proportional to vehicle weight and effectively independent of speed. For car tyres, the rolling resistance coefficient (r) ranges from 0.013 to 0.018, but for fork-lift truck tyres, higher values of 0.02 (for pneumatic tyres) to 0.03 (for solid rubber tyres) are appropriate.

Thus the simplified equation of motion for a truck on a level surface is:

$$F = ma + mgr \qquad (1)$$

where

F = propulsive force (at the wheels)
m = vehicle mass
a = vehicle acceleration
g = gravitational acceleration
r = tyre rolling resistance coefficient

If $(a + gr) < 0$, then $F < 0$ and the truck is braking.

We can integrate Equation (1) with respect to distance (s) to get the energy equation:

$$\int F\,ds = m\int a\,ds + mgr\int ds$$

$$\Big[E\Big]_1^2 = m\Big[\frac{V^2}{2}\Big]_1^2 + mgr\Big[s\Big]_1^2 \qquad (2)$$

where

E = propulsion energy
V = vehicle velocity
s = distance travelled

In other words, the propulsion energy required between any two points (1 and 2) in a driving cycle is $(E_2 - E_1)$, and is the sum of the kinetic energy <u>increase</u> $(\frac{1}{2}mV_2^2 - \frac{1}{2}mV_1^2)$ and the rolling resistance loss $(mgr(s_2 - s_1))$.

If the vehicle is braking, $(E_2 - E_1)$ is negative, and this energy is available for recovery. In an ideal vehicle with a 100% efficient regenerative transmission, the net energy consumption for any driving cycle (beginning and ending with $V = 0$) is therefore equal to the total tyre rolling resistance loss. Let us define this as the Net Propulsion Energy (NPE):

$$NPE = mgrs$$

Note that this depends <u>only</u> on the total distance travelled (s) and <u>not</u> on the details of the driving cycle.

If the transmission is incapable of recovering braking energy, then the total energy consumption will be the sum of all the positive increments of propulsion energy.

If the vehicle speeds are V_1, V_2, V_3 etc. at each driving/braking transition, and U_1, U_2, U_3 etc. at each braking/driving transition, (see Figure 1) then it can be shown that the Gross Propulsion Energy is:

$$GPE = (\tfrac{1}{2}mV^2 - \tfrac{1}{2}mU^2) + mgrs_d$$

where

s_d = total distance travelled during driving (i.e. accelerating, cruising or coasting but not braking).

Thus the GPE can be easily estimated by examining a speed - distance record of the driving cycle, preferably with a separate indication of the start of braking, although this is usually obvious on the speed - distance trace.

Note that the GPE is a function of the driving cycle, and will increase with the severity of the cycle.

We can now define the Percentage of Energy (theoretically) Recoverable as

$$\%ER = \frac{\text{Gross Energy} - \text{Net Energy}}{\text{Gross Energy}} \ (\%)$$

%ER is primarily a function of the driving cycle rather than the vehicle characteristics and it is a very useful indication of the proportion of propulsion energy available for recovery, and hence of the viability of regenerative braking system.

For private cars in urban or suburban conditions, %ER is around 30 - 40% (1), (2), while for city buses, %ER is estimated to be 55% (3). For fork-lift trucks, there are few published standard driving cycles, but analysis of one published duty cycle (4) gives an estimated value of %ER of 75% for a diesel truck.

Such a high value of %ER is not really surprising when one considers the extremely severe nature of fork-lift truck operation, where the truck tends to be driven hard to maximise productivity. The limited acceleration and the need to negotiate tight corners at reduced speed leads to a driving pattern

© IMechE 1985 C275/85

consisting almost entirely of continuous acceleration interspersed with frequent braking, to a reduced speed rather than a halt. It is the frequent reductions of speed for cornering which lead to the high %ER figure, and a simplified analysis based solely on the distance between stops (i.e. assuming one acceleration and one deceleration between each complete stop) will grossly underestimate the %ER.

The fork-lift truck is therefore a much more attractive proposition for a regenerative transmission than normal road vehicles. Not only is the %ER high, but intensive industrial use gives a high ratio of energy costs to capital costs, further improving the economics. The additional weight of a (hydrostatic) regenerative system is also no disadvantage on a counterbalance truck.

3 ELECTRICAL vs HYDRAULIC REGENERATION

3.1 Electrical Regeneration

Firstly, electrical regeneration is only applicable to battery electric trucks. This immediately introduces two drawbacks:

- electrical energy costs are perceived by the user as low (although the real cost, including battery depreciation and charging cost, is much higher).

- electric trucks have a limited driving performance, and the %ER available is therefore much lower. Lift and auxiliaries consume relatively more energy, and thus the regenerable energy is a smaller percentage of the total.

The regeneration efficiency achievable is also severely limited, primarily because

- the charge/discharge cycle efficiency of a lead-acid battery at very high currents is well under 50%

- motor and controller efficiencies under conditions of hard braking and acceleration are also low (typically 75%).

The overall regeneration efficiency is therefore under 30%. If this is coupled with a relatively low %ER of perhaps 50%, and the fact that drive energy in an electric truck may be only 50% of the total, the overall energy savings are a disappointing 10% or so.

There are also question marks about the effects of the high charging currents during regeneration on battery life and efficiency.

However, electrical regeneration is very easy and cheap to implement especially on a truck with electronic speed control. Thus despite the very modest savings, it is a feature offered by several manufacturers.

3.2 Hydraulic Regeneration

In contrast to electrical systems, the characteristics of hydraulic systems are ideally suited to the regeneration of small amounts of energy at high powers. Energy can be stored in a hydraulic accumulator, and while the energy density (i.e. energy per unit mass) is low, the amount of energy to be stored is also low. The power density of the system is very high, and the efficiencies of hydraulic components at high powers are high (pumping 90%, motoring 85%, accumulator >95%. A cycle efficiency of 70% is easily achieved.

As well as more efficient recovery of braking energy, a hydraulic system gives assisted acceleration with lower peak battery currents, giving a significant improvement in battery efficiency. Alternatively, such a system on an i.c. engined truck would give improved acceleration, or allow a smaller engine to be used.

Various configurations of such hydrostatic regenerative systems are possible, applicable to electric trucks or i.c. engined trucks, and designed to achieve the appropriate combination of energy saving, improved performance or reduced cost in each case.

Three concepts are outlined in Section 4.

4 THREE CONCEPTS FOR REGENERATIVE HYDROSTATIC TRANSMISSIONS

4.1 Concept 1: Parallel Regenerative Hydrostatic System for Electric Fork Lift Trucks

This system uses a hydraulic pump/motor connected to the drive wheels (either directly or through gearing) and in parallel with the normal electric motor drive system. This pump/motor is connected hydraulically to a small accumulator, sufficient to store the kinetic energy recoverable from one deceleration from full speed. The function of this system is solely to recover braking energy, store it in the accumulator, and regenerate it to assist the acceleration of the vehicle. When cruising at or near full speed, the accumulator would normally be empty and the hydraulic system would be inactive.

If a variable displacement pump/motor is used, this system would be very simple in concept and easy to control. The basic hydraulic circuit is shown in Figure 2. The variable displacement pump/motor must be capable of stroking in both directions, and rotating in both directions. As one port is always the high pressure port, stroking the displacement in one direction will always apply the same direction of torque to the output shaft, regardless of the direction of rotation. Thus stroking in one direction might apply a braking torque for forward motion, which would also be an accelerating torque in reverse. With the simplest arrangement of components, actuation of the pump stroke to cause braking in forward motion would slow the vehicle to a halt, and immediately accelerate it backwards. Some logic function is required to control this behaviour, such as the pilot operated check valve shown. Other components, such as a relief valve and a by-pass (or preferably automatic de-stroking) to cope with an empty accumulator, are also required.

This system is straightforward, and offers full, progressive control of both accelerating and braking torques. However, variable displacement units are expensive, run at high speeds and hence require gearing, and are less readily available in the very small sizes (5 - 10 cc/rev) required for the smaller sizes of truck.

The alternative to a variable displacement unit is a fixed displacement pump/motor (which could be a gear, vane, gerotor or piston unit) and a suitable directional control valve. (See Figure 3).

Such a system is less flexible than the variable displacement configuration, as the output torque is not independently controllable except by throttling. However, this may be less of a disadvantage than it seems at first sight. Fine control is still possible (by throttling and/or the use of the electric drive) with a certain loss of efficiency. Such fine control may seldom be necessary, except at very low speeds where the energy losses are negligible.

Such a system is therefore very promising, and the main technical problem is the design of a valve block to give all the required functions in one compact unit.

4.2 Concept 2: Series Regenerative Hydrostatic System for Propulsion of Diesel Fork-Lift Trucks

The basic hydraulic circuit for this system is shown in Figure 4. A conventional hydrostatic transmission circuit is modified by the addition of an accumulator, and the replacement of the usual fixed displacement motors by variable displacement motor/pumps (1 or 2).

Output torque (and hence acceleration and deceleration) are controlled by varying the displacement of the motor/pump(s), independently of engine and input pump settings. Braking energy is recoverable in a fully controllable manner, and because peak power demands are supplied by the accumulator, a significantly smaller engine and pump may be used for the same performance (or better performance achieved with the same engine size).

4.3 Concept 3: Fully Integrated Hydraulic System for Drive and Lift using Accumulator for Energy Storage

This concept takes the ideas of Concept 2 a stage further. The lift (and steering and auxiliaries) are fed by the same high pressure supply, stored in a (relatively large) accumulator, and supplied by a (relatively small) engine and pump. The propulsion system is similar to Concept 2, with the same advantages. Oil supplied for lifting must be throttled to the pressure required by the load. This means that all lifts require the same energy as lifting a full load, and is therefore inefficient. However, the losses may be a small part of the total, and are probably outweighed by savings from regenerative braking. More importantly, the performance of such a system will be extremely high.

Lift speeds (loaded or unloaded) will not be limited by engine power, and will be independent of drive (acceleration) demands. The engine and pump (which can probably be of fixed displacement) can be small, as they supply only the mean power, not the peaks.

The advantages of this system are therefore:

- somewhat reduced fuel consumption
- unlimited lift speeds (loaded or unloaded)
- excellent acceleration performance
- no interation between lift and drive systems
- reduced engine size
- reduced pollution and noise because of more uniform engine operation.

5 CONCLUSIONS

Regeneration of braking energy in road vehicles is a highly desirable but, at present, economically dubious goal. The counterbalance fork-lift truck, by virtue of its necessarily massive construction and demanding duty cycle, offers a very attractive application for regeneration, particularly as an effective system can also offer improved performance and hence productivity.

Electrical regeneration is easy but ineffective. This is primarily due to the low efficiency of a lead-acid battery, and also because of the poor performance of electric trucks allows a much smaller proportion of the total energy to be available for regeneration.

Hydrostatic drive units have characteristics much better matched to the requirements of a fork-lift truck regenerative transmission, where the powers are high but the energy to be stored is low. A hydrostatic system can offer not only significant energy savings, but an improvement in performance and productivity.

REFERENCES

(1) DUNN, H S and WOJCIECHOWSKI, PH (1972), "High Pressure Hydraulic Hybrid with Regenerative Braking", Paper 729145, Proc. 7th IECEC, San Diego, September 1972 pp 989 - 995.

(2) MACARTHUR, D S (1977) "Energy-storing Hydrostatic Transmission for Urban Vehicles", PhD Thesis, Engineering Department, University of Cambridge.

(3) CUNNINGHAM, S U and ABBOT, R J (1974) "Some means of reducing energy consumption by employing hydrostatics", Paper 14, Conf. on "R & D Routes to More Effective Utilisation of Energy", National Engineering Laboratory, East Kilbride, 10 - 11 Sept 1974.

(4) ABELS, T (1971) "How can the Operating Speed of Fork-Lift Trucks be Compared Objectively", Deutsche Hebe und Fordertechnik, 3/17 pp 57 - 60 (in German).

© IMechE 1985 C275/85

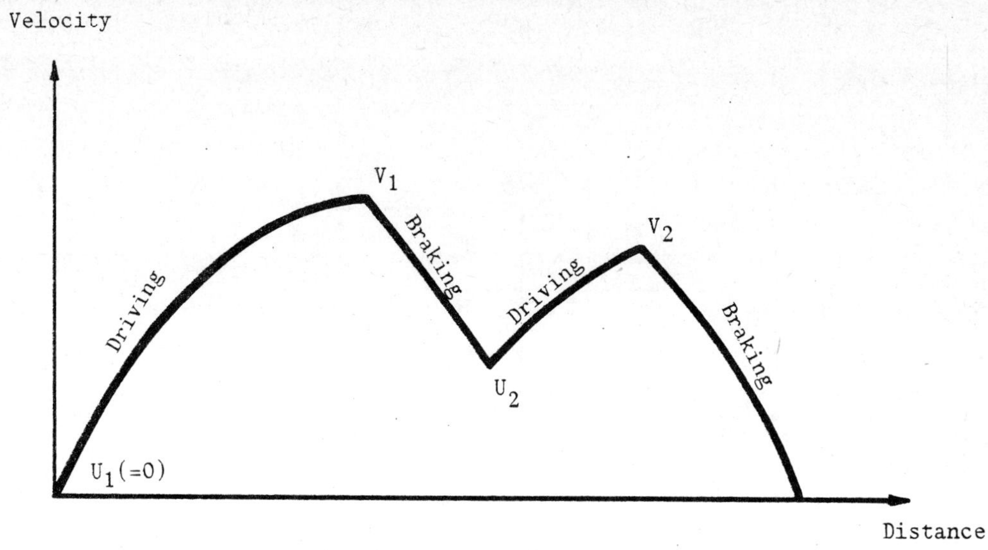

$$\text{Total Acceleration Energy} = \tfrac{1}{2} mV_1^2 - \tfrac{1}{2} mU_1^2 + \tfrac{1}{2} mV_2^2 - \tfrac{1}{2} mU_2^2 + \ldots\ldots$$

$$= \tfrac{1}{2} m \left(\Sigma V^2 - \Sigma U^2 \right)$$

Fig 1 Calculation of gross propulsion energy

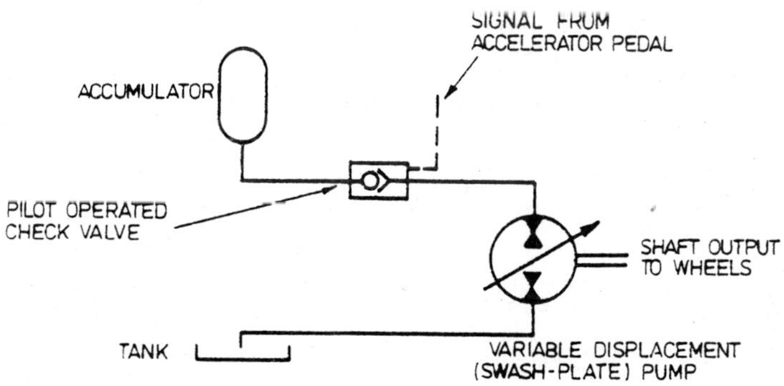

Fig 2 Concept 1a — variable displacement pump

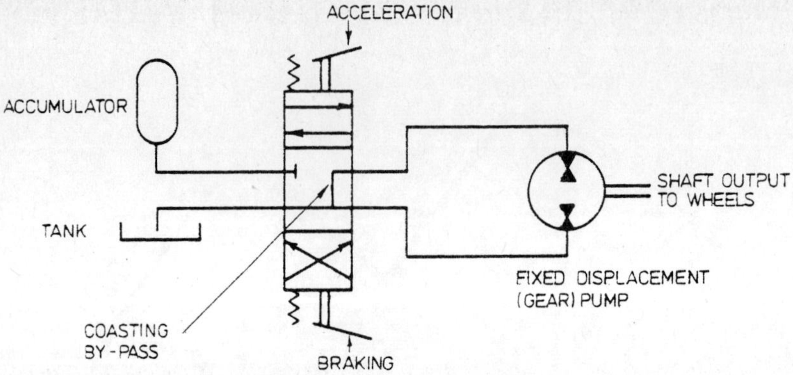

Fig 3 Concept 1b – fixed displacement pump

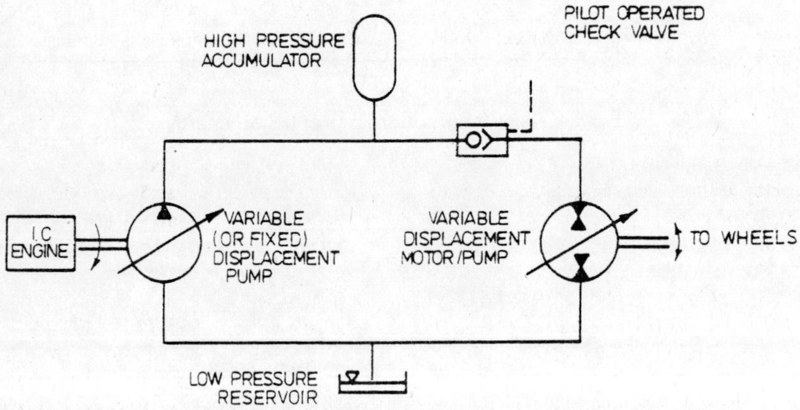

Fig 4 Concept 2 – series regenerative hydrostatic transmission

© IMechE 1985 C275/85

C272/85

A hybrid-electric power system with hydrostatic transmission

J F WATSON, MSc, CEng, MIMechE and A HOOPER, BSc, MSc
University College, Swansea, West Glamorgan

Synopsis

A hybrid electric power system incorporating a hydrostatic transmission is being developed to meet the requirements of a public transport vehicle for use in heavily congested city environments.

This paper describes the design and predicted performance analysis of the system and includes preliminary results from a programme of tests which are being carried out on a full scale test rig.

1. INTRODUCTION

Transport authorities responsible for public transportation in the inner areas of many of the larger cities of the world are becoming increasingly aware of the need for vehicles designed specifically for continuous operation in heavily congested slow moving traffic. This has resulted in both the development of the conventional I.C. engine/gearbox power system for this type of application and also examination of alternative power system concepts. The basic power and transmission system for this type of application must be capable of meeting a continuously varying power and speed demand with frequent stop/start operation; provide good acceleration from stationary with "jerk-free" operation and minimize driver fatigue. Typical operating requirements for such vehicles are:-

Maximum speed (during normal service)	8.9m/sec
Acceleration	0-6.5m/sec in 12 secs.
Average speed	3.87m/sec.
Frequency of stop/starts	4-6 per kilometre
Time stationary	20% to 30%

To meet these requirements, using a conventional power system comprising an I.C. engine and stepped ratio gear box, the engine has to be continuously rated to meet the transient peak power demands, while operating for most of the time at low power output and idling conditions. Fig. 1 shows a typical diesel engine characteristic indicating the high SFC associated with operation at these low power levels.

A hybrid electric power system is being developed specifically to meet the requirements of this type of vehicle. The power unit comprises a diesel engine (operating over a small speed range) rated to meet only the significantly lower power demand associated with typical steady speed operation, with an electrical system providing the additional power necessary to meet the transient peak demand during periods of acceleration. To enable the engine to operate at nominally constant speed over the full vehicle speed range a hydrostatic transmission is incorporated to give infinitely variable transmission and good vehicle driveability. The system also enables full advantage to be taken of electrical regenerative braking during vehicle deceleration.

2. POWER SYSTEM CONCEPT

The power and transmission system is illustrated in Fig. 2. The two power units are mechanically linked by means of a belt drive and both units drive an hydraulic pump through a pneumatically operated clutch provided for starting purposes. There is no separate electric motor controller, therefore the relative proportions of hydraulic pump power provided by the diesel engine and electrical system is determined by the matching of their operating characteristics.

The power sharing is based on the diesel engine providing all of the power for steady state cruising, with zero gradient. The characteristics of the engine and electrical system are then optimally matched such that during periods of acceleration (peak power demand), the natural reduction of engine speed associated with the engine governor characteristic results in a progressive increase in electrical power. After the period of acceleration, the power demand reduces and as engine speed increases, the electrical power reduces and the engine takes over the full power demand.

During vehicle deceleration the hydraulic pump is "motored" by the vehicle inertia through the hydraulic motor and tends to increase engine and electric motor speed. As the speed increases the diesel engine governor shuts off all fuel to the engine while the electric motors automatically regenerate power to the batteries.

Further battery recharging is carried out while the vehicle is stationary when the engine is reduced to half speed and the electric motors operate in a parallel/series mode.

The successful design of a hybrid power system is critically dependant on the selection of the nominal rating of the system components and the matching of their characteristics to achieve an overall system characteristic which in turn matches the load characteristics associated with the vehicle.

This investigation is related to a small citybus having a gross weight of 5000Kgs. The nominal ratings of each component were selected to meet the requirements of this particular vehicle.

3. SYSTEM COMPONENTS

3.1. I.C. Engine

Analysis of the vehicle power requirements indicated that the power required for typical constant speed operation was 28KW while the peak power required was approximately 75KW. The engine is a 2.5 litre indirect injection diesel engine having a nominal rating of 45KW at 3000rpm. The engine, having a performance characteristic shown in Fig. 3, is fitted with a mechanical governor having a speed droop of 12% from no load to full load. The engine governor is set by means of a mechanical actuator to either of two positions for "power" or "charge". The "power" setting determines the vehicle driving conditions at which electrical power is drawn from the battery to supplement the engine. The charge setting determines the rate of battery recharge available and allows the engine to operate at near half speed while the vehicle is stationary.

3.2 Electric Motors

The electric motors used are multiple permanent magnet light weight disc armature motors having operating characteristics shown in Fig. 4. These motors have been specially developed for hybrid vehicle application.

3.3 Batteries

The hybrid system concept is based on the batteries being required to supply energy for transient power during vehicle acceleration and other periods of high power demand only and, as the duty cycle of the vehicle involves lengthy periods when the vehicle is stationary when recharging takes place, the batteries will not undergo continuous deep discharge cycles. Two banks of 12 volt, heavy duty, lead acid batteries are used each having a nominal capacity of 90AH at the 5 hour rate. The battery characteristics are shown in Fig. 4 with the electric motor characteristics and illustrate the battery voltage drop associated with high currents, particularly as level of battery discharge increases.

3.4 Hydrostatic Transmission

The hydrostatic transmission consists of an axial piston variable displacement pump and motor capable of operating up to a maximum hydraulic pressure of 350 bar. Direct drives were required between the engine and pump and between the motor and standard vehicle rear axle input. Standard units are used which are capable of transmitting the peak power requirement and matching the input and output speeds previously referred to. The characteristic for the transmission is shown in Fig. 5.

3.5 Control System

The power system is controlled from an electrical potentiometer, attached directly to the accelerator pedal, which controls the hydraulic transmission through a relatively simple controller.

In the "foot off" position of the pedal the engine is either idling or charging, depending on the current state of battery charge, with the transmission in neutral. Initial movement of the pedal sets the engine to the "power" condition and further movement produces a linear increase in vehicle speed by controlling the swashplate angles of the hydraulic pump and motor. The first 40% of pedal movement increases the swashplate angle on the pump and the final 60% of movement decreases the swashplate angle on the motor.

Rapid changes in pedal position could produce excessive vehicle acceleration and deceleration combined with large electrical currents. The controller therefore incorporates a means of regulating the ramp rates of the control signal to the hydraulic units to limit both acceleration and deceleration, combined with an overload protection for both power and regenerative conditions

4. MATCHING OF COMPONENT CHARACTERISTICS

The main parameters controlling the matching of the individual power system components are the relative speed of the engine and electric motors, battery voltage and battery capacity.

The "power" governor setting for the engine was chosen such that when the engine is providing the total power associated with a vehicle speed of 9m/sec the engine speed is 2600rpm. At this speed the electric motors are required to run at their no load speed. Matching was achieved by selecting a no-load battery voltage of 84 volts (7x12 volt batteries per bank) which corresponded to a no-load motor speed of 3550rpm giving a speed ratio between the engine and electric motors of 1.35. The predicted overall characteristics for the combined engine/electric motor/battery power system are shown in Fig. 6.

5. VEHICLE APPLICATION

A vehicle operating in dense traffic situations with frequent stop/starts requires good acceleration at low speed. Comparison of the overall power system characteristics with

© IMechE 1985 C272/85

typical vehicle characteristics superimposed on Fig. 5 indicates that the hydrostatic transmission meets this requirement with the electrical system providing approximately 60% of the total power. The matching of the components ensures that the engine takes over the full power demand for steady speed operation up to the required operational speed of 8.9m/sec.

6. SYSTEM TESTS

A full scale test rig has been constructed which can be loaded by combinations of hydraulic, friction and inertia loads to simulate any steady state or transient vehicle driving condition. Tests under steady state conditions have shown that the system is stable at any operating condition and the simple control concept employed has been shown to be satisfactory. Typical preliminary test results are shown in Table 1. The results confirm the overall power system and transmission characteristics except for the power produced by the electrical system during peak power conditions which is less than predicted. In actual vehicle driving conditions peak power demand will only last for several seconds compared with the longer periods associated with steady state testing. Dynamic test results will therefore be much more meaningful. The low electrical charging efficiency of 55% is due to the motor/generator units operating at low power levels. It is intended to improve this by using only one disc armature motor for charging.

Prior to further detailed optimisation of the system a series of dynamic tests are currently being carried out simulating vehicle acceleration, to be followed by simulated duty cycle testing under computer control. Preliminary dynamic testing only has been carried out so far, however a typical set of test results are shown in Fig. 7.

For this test the test rig flywheel inertia simulated the vehicle inertia (without passengers) and the controller "ramp rate" was selected to give simulated vehicle acceleration of 0 to 8m/sec in 12 seconds. This rate of acceleration was achieved with the batteries fully charged however, with batteries partially discharged, acceleration was significantly reduced. Battery state of charge has been shown to be an important parameter in determining the ratio of electrical to engine power during any transient operation. Direct measurement of state of charge is difficult. However a process for continuously assessing this is being incorporated into the test rig and it is expected that a similar device will be required in actual vehicles. Preliminary results suggest that the battery state of charge must not be allowed to drop below 30% for good vehicle dynamic performance.

7. COMMENTS AND CONCLUSIONS

The system under development comprises a unique combination of mechanical, electrical and hydraulic components.

The development is related to a specific application. However it is providing a useful insight into the way in which the characteristics of such components can be matched to provide a particular overall characteristic. As with the system being developed, it is then required to assess whether the improved operation and reduction in operating costs offsets the increased capital cost and complexity which obviously is associated with any hybrid system.

The system under development has not yet reached the point where sufficient reliable data is available for such an assessment to be made however, it has been shown that such a system can be designed and operated.

Table 1: Typical test rig results from vehicle simulation tests (steady state)

TEST		A	B	C	D	E
Hydraulic Transmission						
System Pressure	bar	–	20	60	130	200
Motor Power	KW	–	3.6	13.4	25	38.7
Motor Speed	R/MIN	–	485	1172	1130	1100
Pump Power	KW	–	4.6	16.1	32.8	50.7
Pump Speed	R/MIN	–	2418	2277	2169	2137
Efficiency	%	–	72	83	76	76
Diesel Engine						
Engine Power	KW	5.5	5.37	14.4	26.7	37
Engine Speed	R/MIN	1472	2418	2277	2169	2137
SFC	g/KWh	397	620	220	210	210
Electrical System						
Motor Power	KW	−3.7	−0.8	1.72	6.1	13.7
Motor Speed	R/MIN	1975	3250	3070	2925	2880
Battery Power	KW	2.8	0	3.83	8.2	17.5
Battery Current	AMPS	30	0	50	107	220
Battery Volts	VOLTS	95.7	84	78.1	77.1	76.5
Electrical Efficency	%	53	–	45	74	78

Simulated Vehicle Driving Condition

A = Stationary-Charging
B = Steady speed 4.4m/sec
C = Steady speed 10.55m/sec
D = Accelerating at 0.2m/sec^2 at 10m/sec.
E = Accelerating at 0.4m/sec^2 at 9.5m/sec.

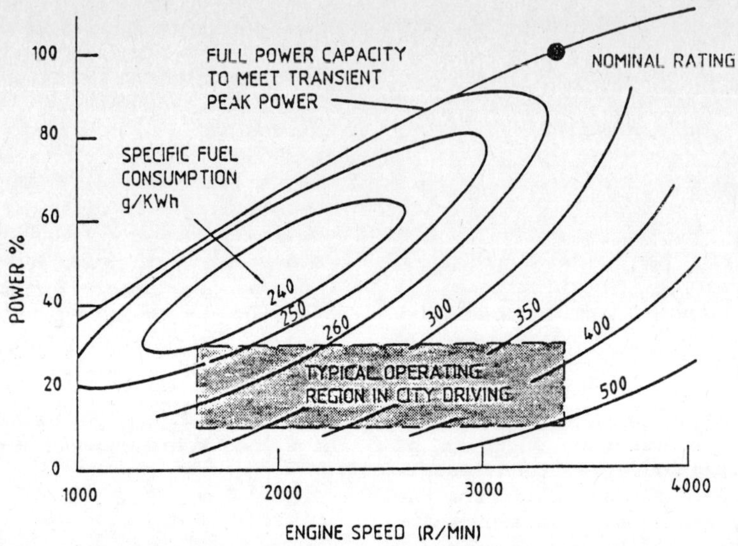

Fig 1 Typical diesel engine operation in city-centre driving (constant vehicle speeds)

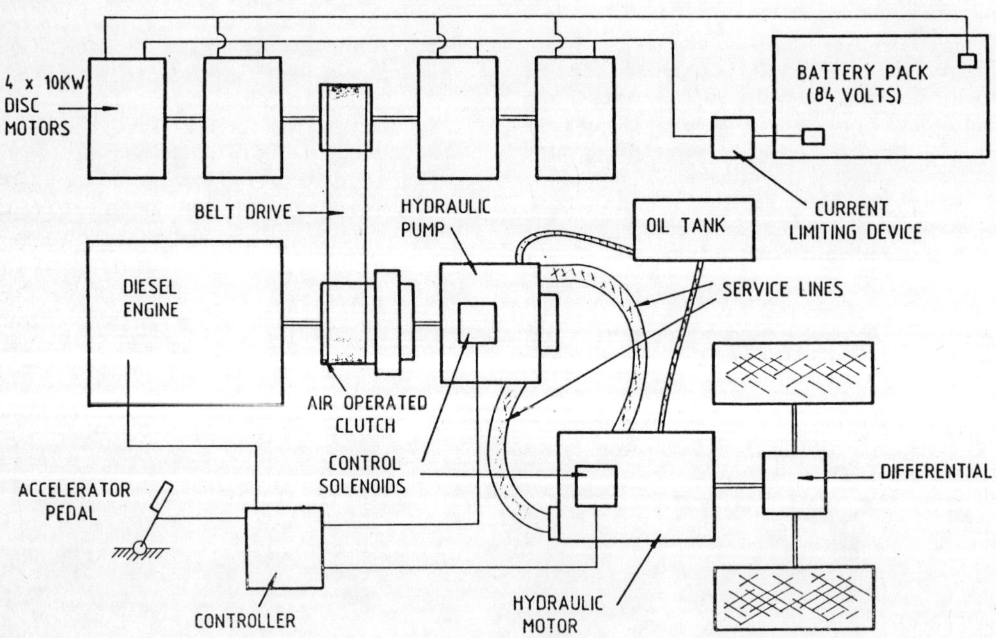

Fig 2 Block diagram of hybrid power system for midi-bus

© IMechE 1985 C272/85

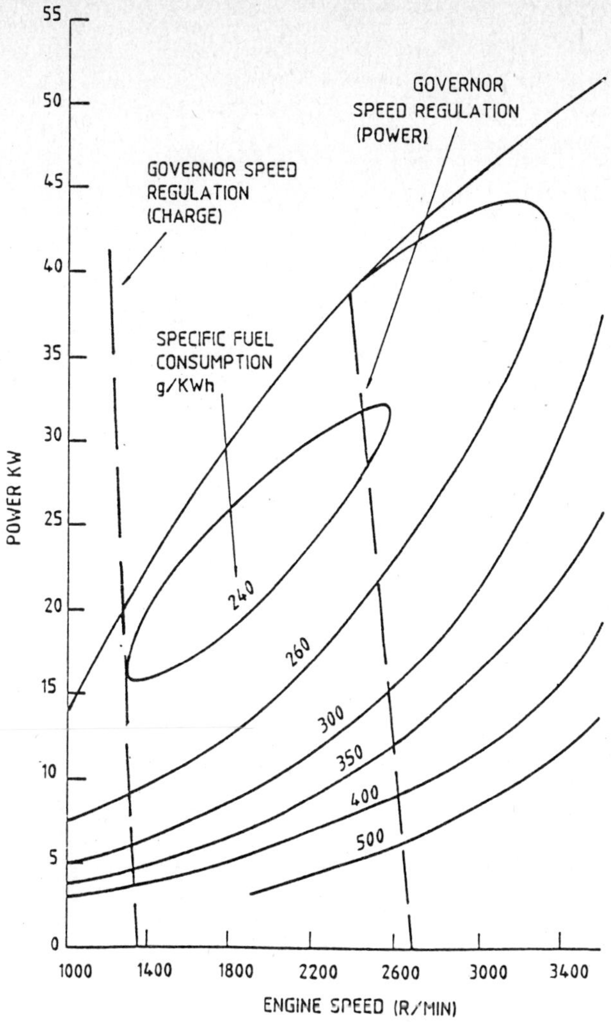

Fig 3 Diesel engine characteristic

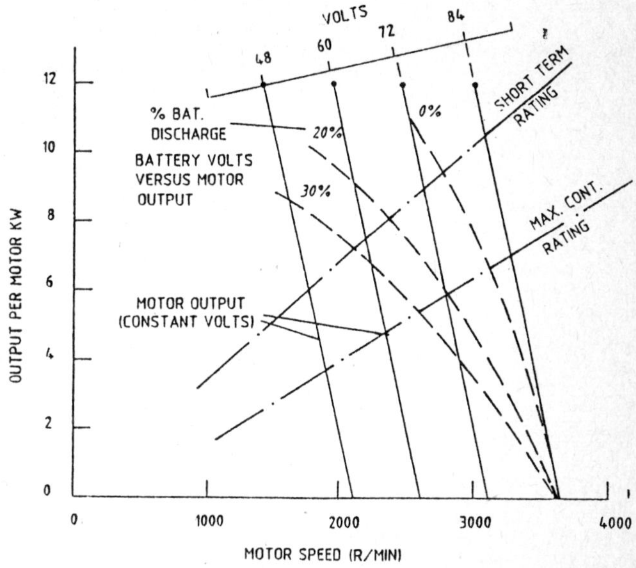

Fig 4 Permanent magnet motor and battery characteristics

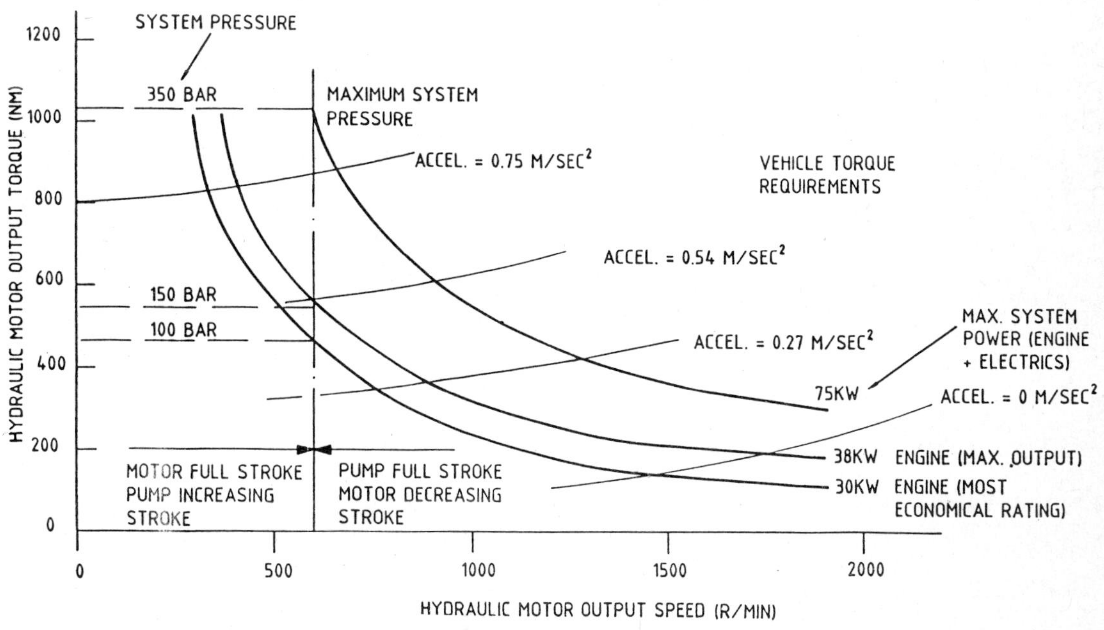

Fig 5 Hydrostatic transmission characteristics

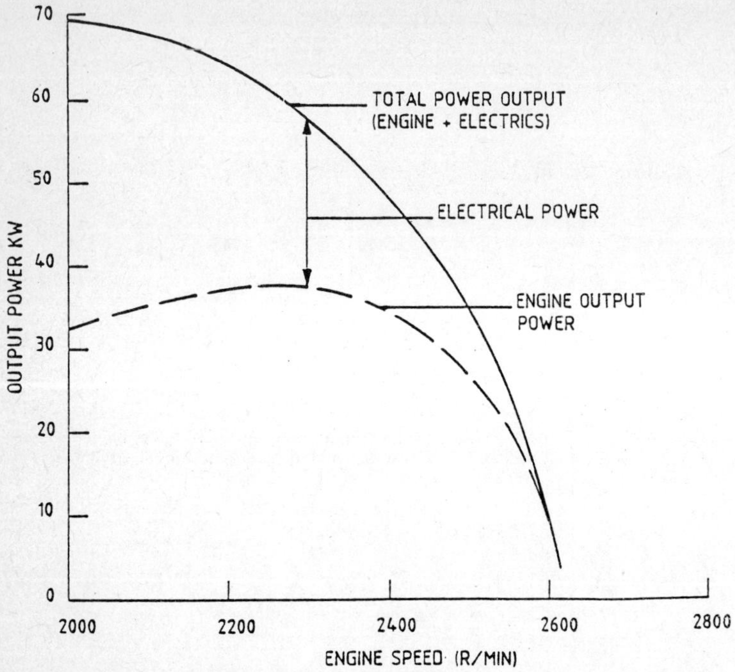

Fig 6 Relative contribution of engine and electrical power to total power output with reduction of engine speed

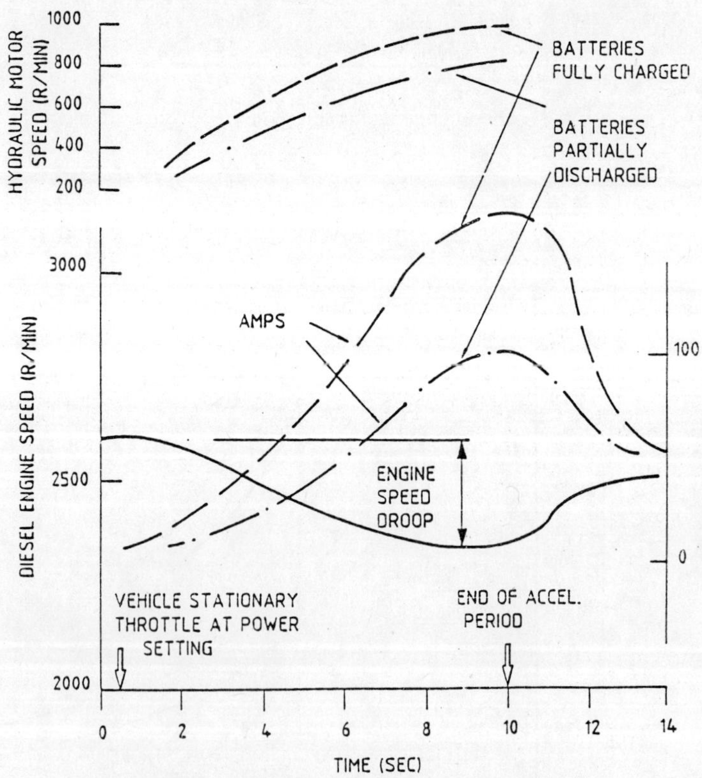

Fig 7 Rig acceleration tests with simulated vehicle inertia

© IMechE 1985 C272/85

C274/85

Electrics and hydraulics on roadheader machinery

B B COTTON
Dosco Overseas Engineering Limited, Newark, Nottinghamshire

SYNOPSIS Roadheading equipment is becoming increasingly important in the development of mineral resources and also on major civil engineering projects throughout the world. This paper discusses the reasons behind the choice of electric or hydraulic drive for the varying requirements of these machines.

1 INTRODUCTION

Dosco Overseas Engineering Limited, part of the Hawker Siddely Group is one of the worlds major suppliers of roadheading type equipment. The major customer is the National Coal Board, but overseas markets account for a substantial part of the business. These include the U.S.A., Canada, China, India, South Africa, Spain, West Germany, Poland, Hungary, Romania, Czechoslovakia.

The equipment is used in the mining and civil engineering industries and takes many forms. A typical example of this is a machine called the TB2000 (see fig 1) which is one of the latest range of machines and is in use in the U.S.A. and Great Britain (NCB).

The machine is electro-hydraulically operated and cuts the strata by movement of two electrically driven cutting heads which are in turn positioned by hydraulic cylinders in the vertical and horizontal planes. The cut debris is then loaded by means of a hydraulically powered gathering and conveying system. The machine has Caterpillar tracks which are driven by hydrostatic motors via reduction gearboxes, the weight of the machine being 80 tonnes.

During the next part of this paper each section of the machine will be explained in detail and the reasoning given for the choice of each particular type of drive.

2 CUTTING BOOM

The cutting boom (fig 2) is normally powered by a 150 kW water cooled electric motor with a synchronous speed of 1800 rev/min U.S.A. and 1500 rev/min U.K. which drives via a spider coupling through a 30/1 epicyclic gearbox giving a cutting head speed of 41 r.p.m. The boom may be utilised to cut rock up to 140 MP compressive strength.

Before outlining the advantages gained from the use of an electric motor in this application it is important to understand the cutting action of the boom. The cutting head on the machine is 1 metre in diameter and 1.1/2 metre long to which are fitted 42 cutting picks which have tungsten carbide inserts. The picks remove the rock by means of a milling action. The speed of cutting varies depending on the type of material to be cut and relies on a hammering effect of the pick rather than the cutting effect as would be seen with a traditional miller.

Initially the power transmission components must be designed to handle the normal running loads plus the torsional shock loads. Torsional shock loads are generally due to the inertia of the system being decelerated when the driven equipment is abruptly slowed or jammed. The magnitude of the torsional shock load is dependent on system inertia, speed change and time.

To achieve the maximum cutting performance through viable rock use is made of the inertia of the system which will create a flywheel effect of 900 times the inertia of the electric motor.

The system inertia is probably the most overlooked item in any design. System inertia acts as a flywheel, increasing the stored energy available to the drive system. However, as recognised by most design engineers, system inertia is a primary cause of damage to power transmission components. The key to a well balanced design is providing sufficient inertia for the system to continue functioning in a non-homogeneous environment without oversizing the motor while limiting the torsional shock loads to a reasonable level.

Figure 3 shows the effect of stopping the cutter boom when fitted with a 1200 rev/min motor; Torque is shown on the vertical axis and stopping time on the horizontal axis. Curve A illustrates the magnitude of the torsional shock loads when a standard coupling is used between the electric motor and gearbox. The full load torque and peak torque levels of the motor are shown for comparison.

On tests recently carried out at Dosco on the MKIIB a smaller single boom machine, the coupling was replaced by a torque transducer. Readings of power input (kW), actual slewing and lifting forces recorded at the cylinders and (by means of strain gauges on the boom) the actual forces at the cutting head were recorded. It should be pointed out that the pressure in slewing and lifting cylinders is of that supplied to the cylinders.

From the results obtained it would be seen that the torque varied from a minimum of -410N m to a maximum of 2680N m. The nominal value being 534N m. During the actual test, fluctuations in torque took place up to 50 times per second. When comparing the actual forces being applied to the machine against the forces the machine is generating, it can be seen that there are large fluctuations in the boom torque and lifting and slew forces whilst the power input (kW) and cylinder pressures (bar) remain relatively stable. This variation is caused by the hammering effect of the picks whilst cutting the strata, enhanced by the inertia of the cutting head enabling the machine to generate higher forces than the nominal rating of the machine would suggest. Also, the forces recorded on lift and slew from the strain gauges are higher than the lifting and slew cylinders can generate. The forces generated due to the cutting action of the boom are at such a speed that the relief valves on the cylinders cannot act quickly enough to dissipate the force being generated.

In an effort to limit the forces to an acceptable level, the equipment can be fitted with a torque limiting friction clutch. Curve B on figure 3 traces the torque level with such a clutch applied. It can be seen that the torque is above the standard system torque level for stopping times greater than 0.26s due to the inertia of the clutch which is added to the system. System torque reaches the slip torque of the clutch at point 'S' indicating the clutch will limit any torsional shock load resulting from a 0.34s stop or instantaneous deceleration rate equivalent to a full stop in 0.34. At stopping times (or equivalent deceleration rates) less than 0.26s (point 'X') the clutch functions as a torque limiter. For example, at a stopping time of 0.1s which is equivalent to 1.0 revolution of the motor or 75mm of cutter movement, the result will be a torsional shock load of 4610N m without the clutch or 2980N m with the clutch, a 55 per cent higher peak without the clutch.

The curve to the right of point 'X' indicates that the additional inertia can sustain system torques for a longer period of time when the torque limiting clutch is used. For example, with the clutch, a torque of 2580N m will be sustained for 0.34 and without the clutch 0.26s, an increase of 31 per cent. This time increase for a given torque level is due to the increased inertia or flywheel effect of the clutch. The net result shown on curve B is that the safe level of torque can be transmitted for a longer period of time. The flywheel effect has increased the energy available for breaking through the 'hard spots' without increasing the torque above the safe level defined. A term "torque enhancement" is applied to the clutch flywheel effect. Thus the use of the inertia of the electrical motor is a vital pre-requisite for successful cutting. Trials have been conducted with hydraulic cutting motors on different machines. The motors used varied from the conventional type gear motor used at 1500 rev/min to the slow speed high torque motors used at 500 rev/min and as expected disappointing results were obtained with their use.

The disappointing results obtained from the use of hydraulic motors are due to lack of inertia and therefore stored energy in the system, and as the cutting boom hits hard bands of rock above the normal cutting range of the machine the boom tends to stall rather than break through the hard band.

It is theoretically possible (through the use of a torque limiting clutch and flywheeel) to design a hydraulic motor driven system with the cutting characteristics of an electric motor system (and is a potential area of research and development).

Later in the paper, the effects of the inertia of the motor when the cutter boom is used in civil engineering applications will be discussed.

3 DEBRIS GATHERING AND LOADING ARRANGEMENT

This consists of two spinner type feeders and a centre strand conveyor (see fig 4.) The running speed of the conveyor is 2m/s and it is driven through 25.8/1 reduction gearboxes by two (F150 Volvo Bent Axis) hydraulic motors running at 1500 rev/min with an oil supply of 7.7 l/s @ 138 bar which is teed off giving 3.9 l/s to each motor.

At Dosco both electric and hydraulic conveyor drives are manufactured. During trials on the test site and in the field no discernible difference in performance between either has been noticed.

For the electric drive a fluid

© IMechE 1985 C274/85

coupling is necessary, because if the conveyor jams, damage will occur, and the fluid coupling also limits the starting torque of the motor.

For the hydraulic drive the starting and stopping characteristics of the motor mean that a fluid coupling is not necessary. The lives of the gearboxes when used with electric or hydraulic motors are within acceptable limits of each other and although it is difficult to make absolute judgement because installations vary greatly it would appear that a greater life of gearbox and conveyor chain is being achieved with the hydraulic drive.

Thus the choice of drive (either hydraulic and electric) system for the conveyor could be taken on its own merit when fitted to this TB2000 machine.

During the course of the design of the machine it became increasingly apparent that due to space limitations it would be very difficult to fit an electric drive to the gathering arrangement but, however, the decision to fit a hydraulic drive could not be taken in isolation from the rest of the machine as the hydraulic power pack and oil tank capacity would have to be increased. This had the effect of increasing the machine overall cost quite substantially. Unless the oil destined for other parts of the machine could be utilised, then the cost of fitting a hydraulic drive would be prohibitive.

4 TRACTION

The options for systems available to use to drive the tracks are as follows:

4-1 AC/DC variable speed Motors

This system is most commonly used on USA designed continuous miners. On information obtained from different machines, each AC/DC system is, in principle identical; only the means of controlling the Silicon Controlled Rectifiers (or Thyrstors) differs. A traction system comprises:

1 Transformer to reduce the incoming AC voltage to an acceptable level to the S.C.R.'s and DC motors.

2 Silicon Controlled Rectifiers (SCR) (Power Module).

3 DC Motors.

Two types of control are available with this system, either infinitely variable throughout the speed range or set speeds with controlled increase between speeds. All the DC systems to our knowledge were water cooled.

One advantage of using a DC system over a AC system is that on the DC system loadings can be controlled to prevent shock on the track and gearbox.

4-2 AC Variable Speed Drive Motors

An AC variable speed drive with frequency conversion using Thyristors or Silicon Controlled Rectifiers. This system is complicated and expensive and has not generally been used for this application.

4-3 AC Two Speed Motor

This system requires the following:-

1 Contactor Control Panel.

2 A Twin Wound Motor - this arrangement has two set speeds in forward and reverse and has no acceleration or deceleration controls and whilst possible, has not found favour amongst mining machinery manufacturers.

4-4 Hydraulic Traction System

The hydraulic drive system for traction has found favour with UK roadheading machine manufacturers for the following reasons:-

1 The drive unit is compact.

2 A small gearbox reduction is required using a slow speed high torque motor.

3 The dynamic braking of the system is automatic if an overcentre valve arrangement is fitted directly to the hydraulic motor. Only a small fail safe parking brake is required.

4 The power input to the drive is adjustable by altering the setting of the relief valves. This is a useful facility to cater for increased weights of machines as additional equipment is fitted.

5. Variable speeds can be achieved by utilising oil supply from other hydraulic services.

6. A direct drive onto the gearbox can be usedwith a low speed high torque motor.

7. A machine will not overspeed when fitted with overcentre valves.

The negative points of a hydraulic drive are:-

1. A hydraulic drive is less efficient than an electrical drive.

2. A hydraulic system needs to be checked more regularly on a planned maintenance basis due to there being more components.

3. Potential oil leaks from fittings, etc.

4. A high volume of oil is necessary if a high tracking speed is required. On this particular machine a speed of 0.3 m/s was necessary requiring a flow of 5.2 l/s

5. To be able to fault find with a

hydraulic system, maintenance people with a relatively high degree of training are required.

The costs of a hydraulic drive system and a AC/DC system are comparable, but, there is a higher development cost for the electric drive. Taking account of all the factors listed it was decided to opt for a hydraulic drive and a provisional hydraulic circuit prepared (see fig 5).

SUMMARY

From the preceding text it can be seen that the cutting arrangement will be powered by an electric motor making full use of the system inertia available. The choice of drive for the debris loading arrangement and traction will be made with a view to the overall machine arrangement.

At this point it is important to understand the basis of how the machine will be used. From this it will be possible to understand the logic employed in the design of the hydraulic circuit.

The TB2000 when cutting coal measures strata would normally be expected to cut and load in a two metres high x six metres wide roadway with an advance of six metres in one hour. After cutting and loading the six metres the machine can then be expected to reverse out and then 'travel' a distance of up to 100 metres before it can engage in its next cutting cycle. Therefore, it can be seen that whilst cutting and loading a slow tracking speed is acceptable. After the machine has finished its cutting and loading and is required to move to its next cutting position, the tracking speed needs to be high but cutting and loading are not required. Thus it can be seen that the oil can be used for more than one function. The basic flow to the tracks is 1.1 l/s and to the loading arrangement is 7.7 l/s . For faster movement from one work area to another the 7.7 l/s is split, 5.4 l/s, to boost the track flow (and therefore speed) and 2.2 l/s returned to tank. This could also be used for the tracks if an even higher speed is required. Using much of the same oil for both loading and machine movement enables the oil tank and power pack to be kept to a reasonable size and cost.

The decisions taken on the various forms of drive meant the machine arrangement was as follows:-

1 The two cutting booms were driven by 150 kw electric motors.

2 The remainder of the machine was driven by a 225 kw electric motor driving commercial hydraulic gear pumps producing 15.5 l/s of oil at a pressure of up to 138 bar

5 CIVIL ENGINEERING APPLICATIONS

Over the past decade the use of a boom type tunnelling machine has become a readily accepted method of excavating a tunnel. A conventional roadheader has been used in some applications but, due to ground conditions and in certain cases the finish of the tunnel required, it became necessary to mount the cutting arrangement within a conventional shield. To meet this market, the TM1800 and SB600 (see fig 6) machines, were developed.

The TM1800 and SB600 have since been used in varying arrangements on many tunnelling applications, including the extension of the underground railway system at Heathrow Airport, the driving of the drifts for the conveying and supply systems in the Selby Coalfields and, a four boom arrangement, see fig 7, that was used in a joint venture by Schafir and Mugglin AG/CSC Strassen Und Tiefbau for the excavation of two parallel tunnels 11.4 metres in diameter and 1 km in length for the Rosenburg motorway at St. Gallen, Switzerland.

The life expectancy from the cutting boom arrangement is considerably less than that obtained from the same boom fitted to a conventional roadheader.

The reason for this decrease in life of components is explained by refering back to the discussion of the inertia of the drive to the cutting head. As previously stated the magnitude of the torsional shock loads is dependent on the system inertia speed change. When observing the conventional roadheader cutting, (that being a free standing track machine) it was noted that when cutting exceptionally hard bands of rock of compressive strengths up to 140 mpa the machine tended to lift or slew. Even when cutting in materials of lower compressive strengths a certain degree of machine bounce and machine vibration was experienced.

Thus it can be deduced that the top range of the torsional shock loads are being absorbed by movement and vibration of the machine permitting less sudden stops.

Whilst excavating circular tunnels the boom is confined within the shield which is in turn confined by the material surrounding it. With this highly rigid arrangement machine bounce and vibration is prevented and therefore all the torsional shock is taken within the machine configuration itself. This inability to disperse loads through movement of the machine means several modifications to the cutting arrangement have had to be introduced, some of these being listed below:-

1 Each lifting and slewing cylinder is fitted with a load control valve which is a direct acting differential area poppet type relief valve with externally operated pilot piston and reverse free flow check valve. Also incorporated in this valve is an outer large poppet valve with a twin pilot ratio. The reasons for fitting this

© IMechE 1985 C274/85

valve are twofold; when the cylinder has no oil flow to it and the control valve to the cylinder is in the closed position, a burst hose or leaking connection on the pipework will enable the cylinder to maintain its position, but if a reverse shock load is applied, which happens when the boom is cutting, the boom can relieve through the relief section of the valve which is set at 50 per cent above circuit pressure. During cutting operations within a shield it was found that the relief capacity of the valve had to be increased to prevent bellying of cylinders, caused by very high back pressures. Also, on one particular application when cutting rock of 200 MPA the size of hoses returning to tank have to be increased to accommodate the surges of returning oil.

2 In developing this boom for civil engineering applications, it was found that the front bearing and the bearing arrangement within the epicyclic gearbox needed to be increased in capacity. The shaft driving cutting head had to have an increased material specification and particular attention had to be paid to the removal of all stress raisers.

3 An interesting feature occurred, when the cutting boom was reduced in length for a particular application. This reduction in the length of the cutting head shaft increased the torsional stiffness of the boom and reduced the capacity of the drive shaft to withstand the torsional shock loads that were being imposed. This resulted in the eventual failure of the shaft.

This paper expresses the authors reasoning behind the selection of various drives and the factors behind those decisions.

The views expressed in the paper are those of the author and not necessarily those of Dosco Overseas Engineering Limited.

TABLE 1
Results Obtained During Tests on MKIIB

Boom Torque N m	Lift & Slew Cylinder Pressure Bar	Motor Power kW	Lift Force kN *	Slew Force kN *
-410	62.5	73	53.8 (26.6)	81.7 (34.7)
183	64	73	55.8 (27.2)	78.7 (35.6)
255	64	68	48.8 (27.2)	82.7 (35.6)
268	66.4	83	55.8 (28.3)	86.7 (36.9)
282	57	62	24.9 (24.2)	74.7 (31.7)
310	66.4	70	59.8 (28.3)	80.7 (36.9)
466	69	78	49.8 (29.4)	84.7 (38.4)
502	64	69	40.8 (27.2)	79.7 (35.6)
593	64	65	56.8 (27.2)	74.7 (35.6)
643	65.6	62.5	54.8 (27.9)	109.6 (36.4)
650	61	73	87.8 (26.0)	90.7 (33.9)
720	65.6	62.5	42.8 (27.9)	53.8 (36.4)
792	59	66	48.8 (25.13)	81.7 (32.7)
793	64	70.3	59.8 (27.2)	74.7 (35.6)
2302	65.6	70.3	70.7 (27.9)	93.7 (36.4)
2683	61	70.3	70.7 (26.0)	93.7 (33.9)

The nominal rating for the MKIIB is:-

534	138	83	58.8	76.7

* The values in brackets are the actual loads being applied to the boom by the lift and slew cylinders.

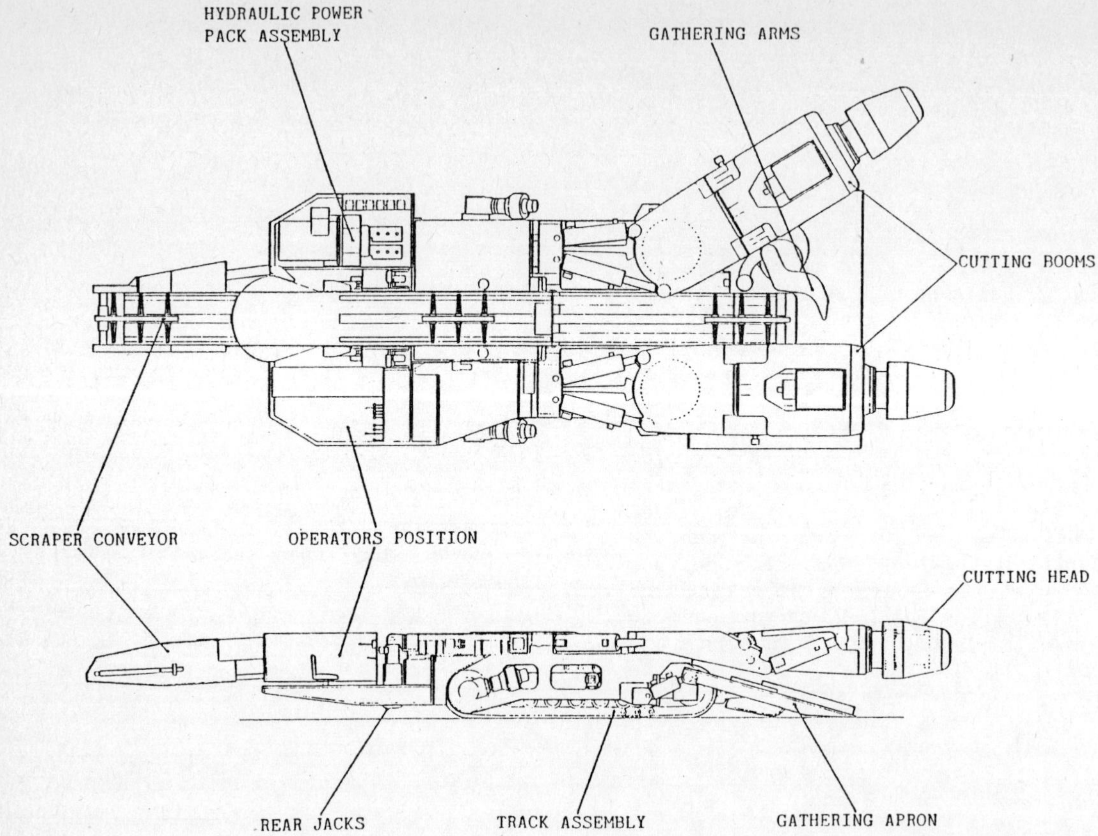

Fig 1 Dosco TB2000 general arrangement

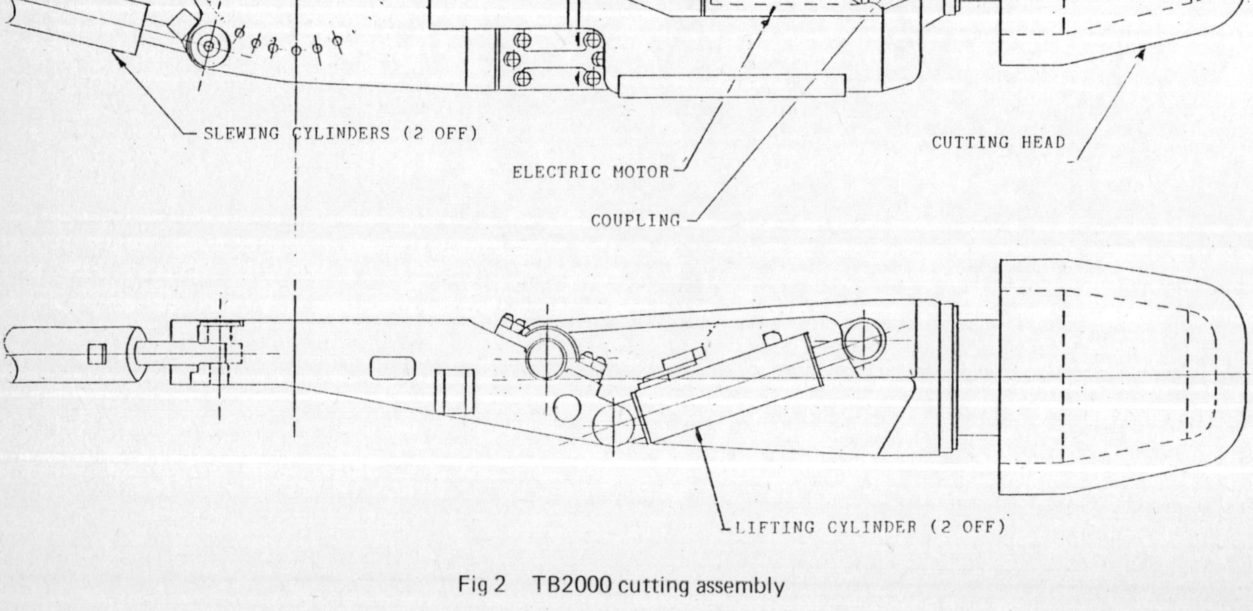

Fig 2 TB2000 cutting assembly

© IMechE 1985 C274/85

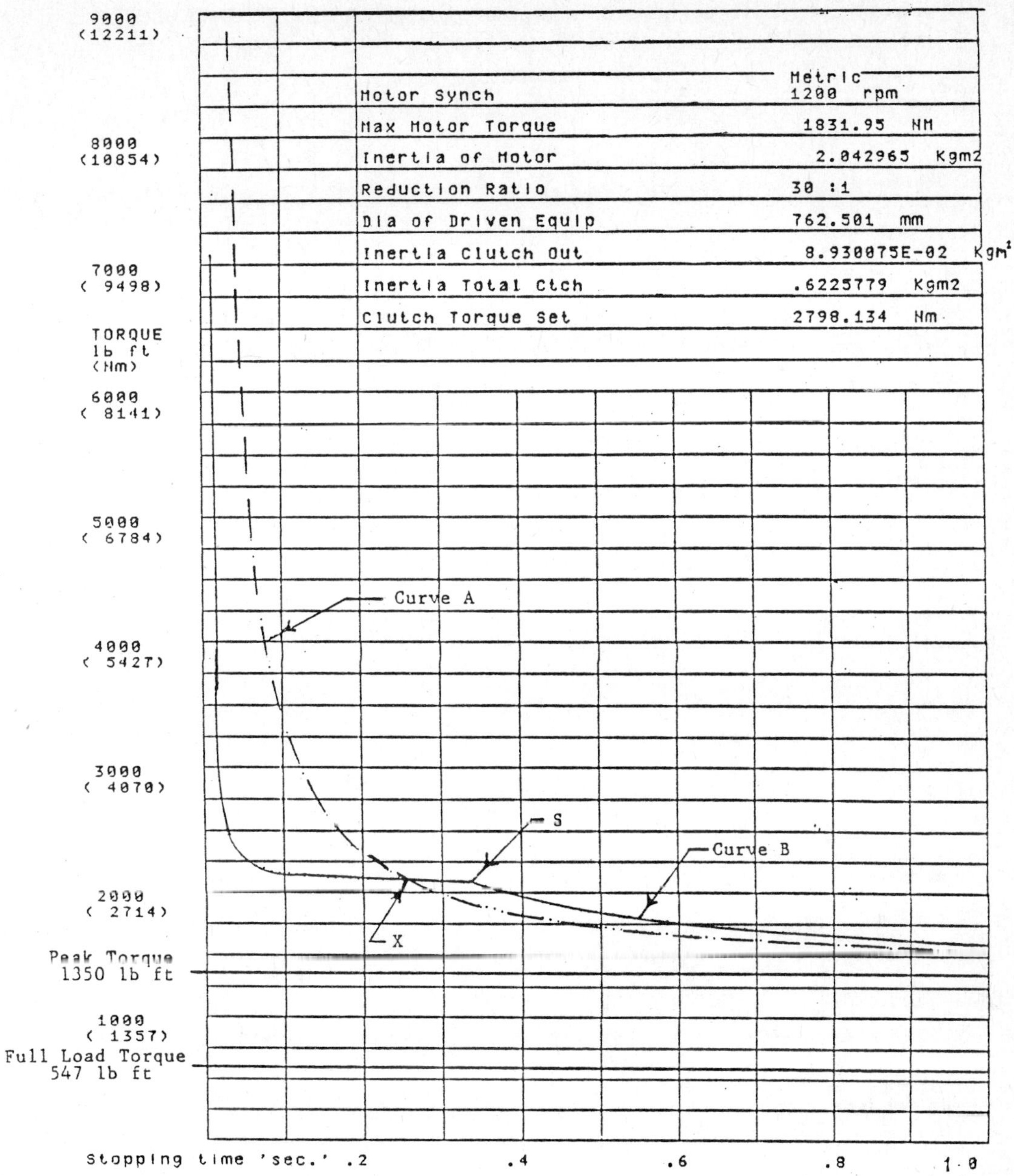

Fig 3 Torque analysis graph for TB2000 cutter head with 1200 r/min motor: curve A – standard coupling; curve B – torque limiting friction clutch

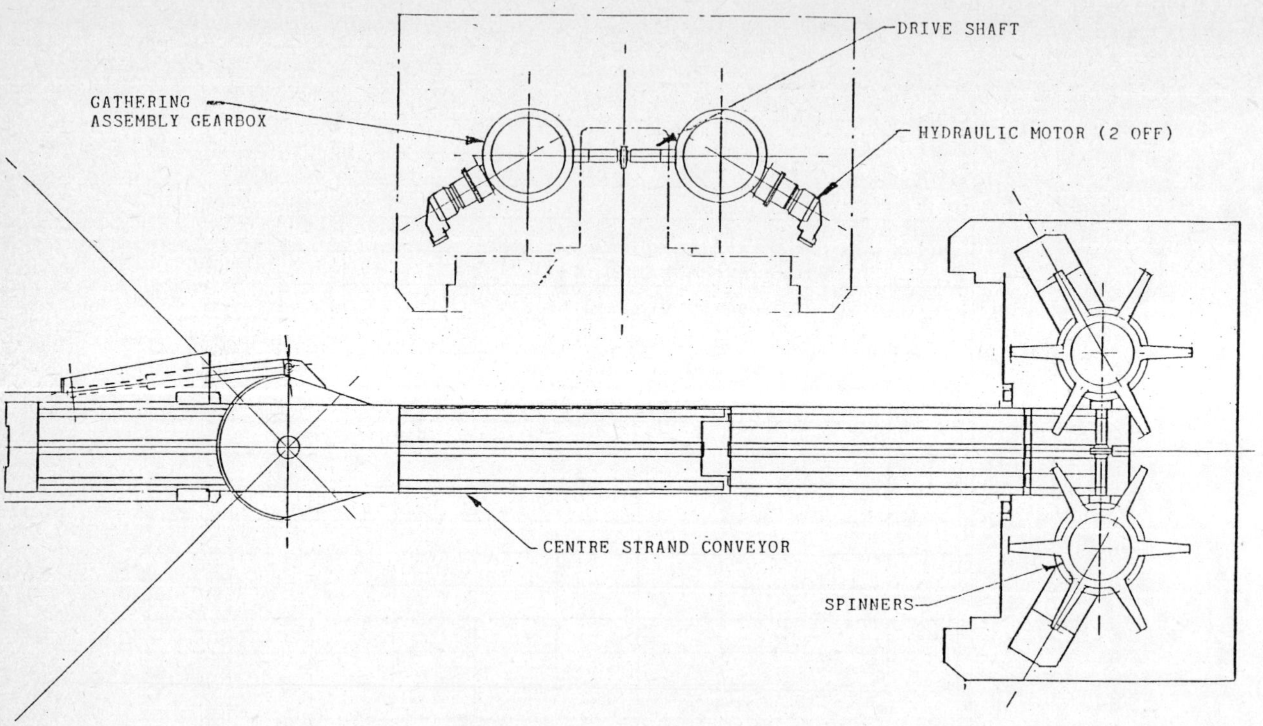

Fig 4 Debris gathering and loading arrangement

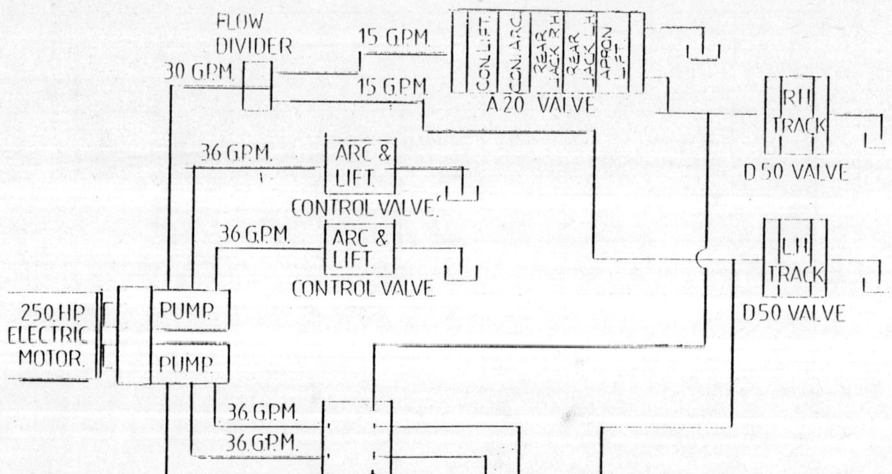

Fig 5 Machine hydraulic arrangement

© IMechE 1985 C274/85

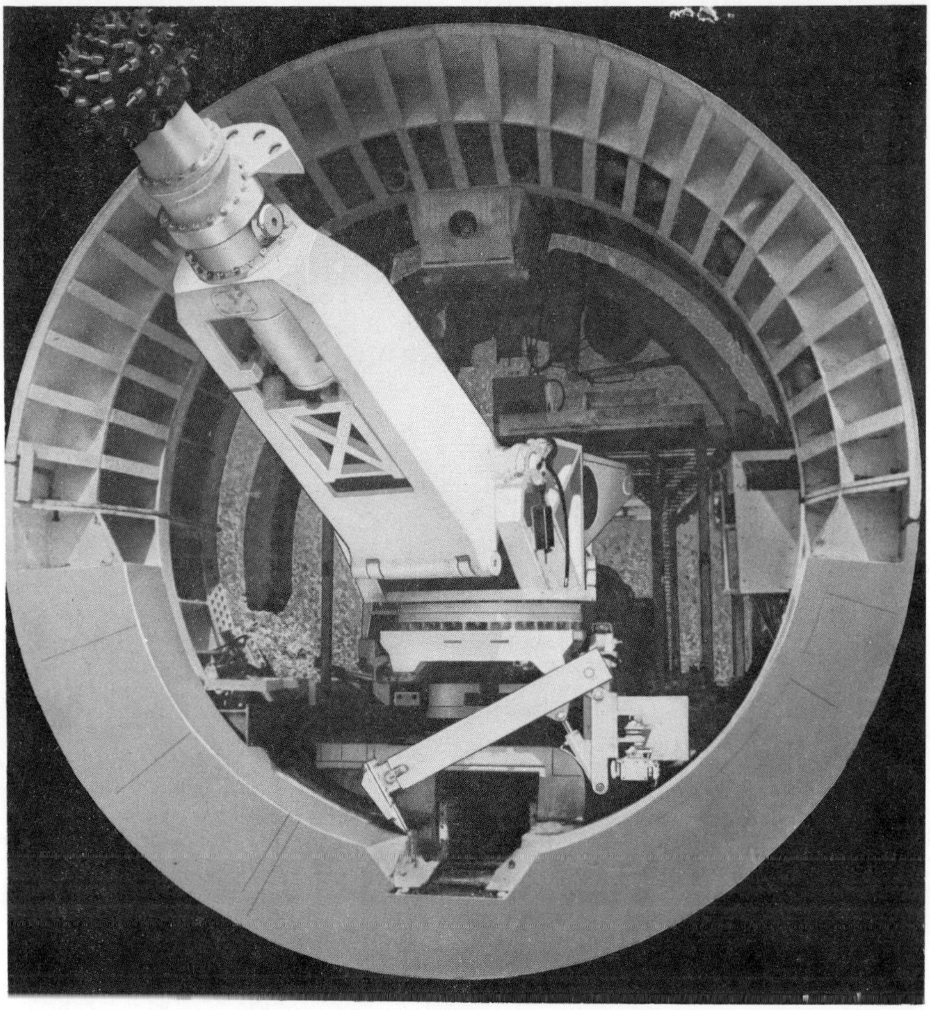

Fig 6 SB600 machine (as used at Don Valley)

Fig 7 Four boom machine for Rossenburg motorway tunnels

© IMechE 1985 C274/85

C268/85

Hydraulic, electrical, pneumatic control – which way to go?

D P ROSE, BTech, CEng, MIMechE and **N JONES**, BSc
Moog Controls Limited, Tewkesbury, Gloucestershire

SYNOPSIS: This paper considers the areas where the various control technologies overlap on the fundamental bases of control power, system dynamics, cost etc. Practical examples from the Aerospace sector of Industry are used to illustrate the advantages and disadvantages of the alternative types of system. These examples are considered in some depth in order that the relative merits of each type can be explained. Much state of the art technology is included. Graphs summarising the operating regime of each technological type are also presented.

1 INTRODUCTION

Control system specification is becoming an evermore difficult task for the Engineer. As the fundamental technological types become more highly developed, applications which at one time would have been reserved for electrohydraulic control systems can now be satisfied by electromechanical systems, and in some cases, electropneumatic systems. For the Engineer, it is important to understand the capabilities of the various technological types when measured against a number of criteria.

A major consideration for any design process is cost and this applies irrespective of the level of technology being considered. Other fundamentals for consideration when designing a control system are the control power and the required overall system dynamics. These and other criteria are considered in the following assessment of aerospace applications from the electrohydraulic, electromechanical and electropneumatic control system fields.

2 ELECTROHYDRAULIC

Traditionally, high performance servomechanisms have been implemented using electrohydraulic systems. Electrohydraulics have, in the past, offered many advantages over other methods of control in terms of power to weight and size ratio. This advantage has been widely exploited especially in the field of aircraft control where the high power density of electrohydraulics has been used and readily allows a high level of redundancy to be built into the system. Recent advances have permitted other technologies to be considered for such applications. However, some of these advances have also led to improvements in electrohydraulic systems.

A typical electrohydraulic flight control system with redundancy is shown in Figure 1. This particular design is for the F-111 aircraft and, as can be seen, contains what appears to be a complex arrangement of valves. The circuit receives three electrical command signals and two hydraulic supplies. All necessary control required to produce an actuator displacement proportional to the input command signal and to detect and act upon failures is contained within the package. Despite its apparent complexity the system is basically two active hydraulic control channels driving a common actuator, plus a third model channel with some failure detection and switching devices.

It is not difficult to conceive that much of the hydromechanical logic and the model channel may be replaced by electrical transducers and electronic logic. However, until recently the bandwidth attainable in practice using digital techniques has been limited. Advances in microprocessor technology now allow system modelling and failure detection to be performed without compromising the high speed digital loop closure necessary in a high performance servosystem.(1) This philosophy is shown schematically in Figure 2. and comparing it to the previous system the potential is obvious.

Prototype units, based upon the above ideas, have been developed. These units comprise a dual tandem actuator, two servovalves with dual spool position transducers, a pair of two stage by pass valves with position switches, four pressure transducers and all electronic hardware and software algorithms to enable the unit to function as a control system. The functions required of the unit are listed below:

* A redundant dual tandem electrohydraulic actuator
* Force summed actuation with active valve null equalization
* Each actuator section controlled by one electrohydraulic servovalve
* Each valve receives dual electrical inputs
* Valve spool position monitored by LVDTs
* Valve monitoring and modelling for failure detection
* Digital loop closure

* Reconfigurable with redundancy management
* Software diagnostic capability and failure recording for maintenance purposes
* Preflight test upon valid command from flight control computer
* Adaptive control by changing control algorithm upon valid command from flight control computer
* Single discrete signalling between computers
* Command and synchronisation from flight control computer

Whilst the units have been seen to operate properly at the time of writing, full evaluation and performance data is not available as the test programme is still underway.

In addition to the advantages gained from advances in digital electronics, electro-hydraulics are also benefitting from the advances in magnet materials, and the availability of samarium cobalt permanent magnets have allowed new types of valves to be developed.

In the past, control valves for high performance systems have been almost exclusively two stage servovalves. There are several designs of such valves on the market and whilst some claim better resistance to contamination than others, all still have the potential of a first stage failure resulting in loss of control. They also have quiescent power requirements. The advantages of these valves is their very high bandwidth and high power amplification.

In order to reduce the chance of a valve failure and also to increase overall system efficiency, single stage designs using samarium cobalt magnets have been developed. These high energy magnets form the basis of a small linear actuator or force motor which directly drives the main spool and this arrangement has given rise to the name Direct Drive Valves. Whilst these valves cannot completely match the very high performance two stage servovalve on bandwidth, their performance is in excess of what most systems require.

The force motor in the valve operates on the same principle as the torque motor in a servovalve. An armature is held in the magnetic flux generated by the permanent magnets, and by applying a bi-polar current to the electrical coils a force is applied to the armature, causing it, and hence the spool, to move. Feedback, to give a spool position proportional to input current, is achieved by either a spring, an LVDT, or a combination of the two. Whilst the concept of this design is not new the availability of high energy product magnets has made it possible to achieve high spool driving forces and acceptable bandwidths with a reasonable drive current level.

The aim in citing the above examples is to illustrate how advancing technology is changing electrohydraulics. It is easily seen that by utilising recent advances it is now more possible than ever to design a high performance,small, lightweight, sophisticated control system.

3 ELECTROMECHANICAL

Advances in magnetic materials and power electronics have increasingly allowed high performance electromechanical servosystems to be offered as an alternative to electrohydraulic systems. This section examines an advanced high performance servoactuator based on brushless servomotor technology.

The availability of samarium cobalt permanent magnets has made it possible to reduce the mass of the magnets required in a brushless motor. This reduced mass leads to designs with lower rotor inertia. Also the increased ability of samarium cobalt to withstand demagnetisation allows higher peak torques to be realised. The combination of these two characteristics results in improved dynamic response in comparison with more traditional designs. In addition power electronics are now widely available to reliably control the commutation at high motor speeds. The nett effect of combining all these advances is the use of a higher motor speed to achieve the required actuator performance from a smaller motor and overall actuator size.

The actuation system to be discussed (2) is a 0.19 KW linear output servo system designed to meet demanding frequency response and duty cycle requirements at minimum size and weight. The system is shown in block diagram format and a sectioned drawing of the actuator is presented (Figure 3). The linear output is achieved by driving a precision ball screw with the brushless motor.

The motor comprises samarium cobalt permanent magnets on the rotor and electrical coils in the stator. The rotating field in the stator is synchronised with the flux established by the permanent magnets using Hall effect transducers to sense the rotor position. The rotary output of the brushless motor drives the precision ball screw through a spur gear reduction. Position feedback from the linear output is provided by an LVDT and a tachometer provides the necessary signal to close an inner velocity loop.

The inner velocity loop provides a convenient means of achieving a stable position loop with good static performance. The effects of magnetic cogging as well as amplifier anomalies are significantly reduced by using this inner integrating loop.

Management of the system is achieved using an electronic controller for a number of tasks. The controller firstly provides the necessary servoelectronics to achieve a stable closed loop position system using the position and velocity signals. The pulse width modulation required to control the motor efficiently is another of the controllers functions and this is transferred to the motor through power switches with flyback diodes. This provides a three phase full wave motor drive. The switches and motor are protected against over heating by current limiting of the output. The controller finally synchronises the P.W.M. output to the stator windings with the permanent magnet flux using the rotor position signal from the Hall effect transducer.

© IMechE 1985 C268/85

The size of the motor and drive train is selected to achieve the load velocity and frequency response duty cycle whilst keeping within the size and weight envelope. Ideally the coupling gear ratio is as high as possible to minimise motor size whilst retaining good frequency response at a suitable current limit.

The frequency response of the system is limited by the voltage and current available at the motor windings and the limits are shown in Figure 4. It can be seen that the response is limited in the range one is normally interested in by the torque available at maximum motor current.

Performance data obtained from the system is given below:

Maximum Output Force (stall)	1.78 kN
Maximum Speed (no load)	0.21 m/Sec @ 22 VDC
Maximum Power Point (1.424 kN @ 0.13 M/Sec	0.186 KW
Stroke	±0.024 m
Resolution	20×10^{-6} m
Stiffness	18.9×10^{6} N/m
Frequency Response	
for ±5% command	90° @ 18Hz
for ±0.8% command	90° @ 38Hz
Continuous Duty	
output force	828 N
output power	0.096kW
Pin to pin distance (retracted)	0.167 m
Actuator Mass	1.2 kg
Supply Voltage	28 vdc
Motor Current Limit	23 amps

Response characteristics are shown in Figure 5. The frequency response figures were obtained by driving an inertia of 0.09 kg m^2 though a 0.08m lever arm with the inertia connected to a load spring of rate 1.8Nm/rad. To obtain the load-velocity profile the spring load was replaced by a hydraulic motor.

The actuator system has been evaluated for both continuous and limited duration duty cycles, the motor operating reliably with a coil temperature of up to 150°C. The following table shows the force-time envelope that can be achieved if the motor windings are to be maintained at temperatures less than 150°C.

Time for coil temperature to reach 150° C

Force	1780N	1240N	
Time to 150° C	95 sec	600 sec	38° C ambient
	40 sec	180 sec	94° C ambient

4 ELECTROPNEUMATIC

A further alternative for high performance servomechanisms is electropneumatics. Such systems have many attractive features, but perhaps the most appealing is the ability to operate from a simple stored energy system while at the same time providing adequate performance at potentially lower cost. Further advantages can be identified as a wide temperature capability, essential for simple servo-electronic controls, and low sensitivity to high acceleration and vibration inputs.

The most obvious disadvantage associated with the use of pneumatics relates to gas compressibility. The high resultant compliance of any pneumatic system manifests itself in the forms of reduced dynamic performance and lower levels of static accuracy (unless friction levels are particularly low). Further disadvantages are the susceptibility of the pneumatic servo system to performance variations with changing supply pressure and changing actuator volume.

What is often needed to make the pneumatic solution viable is a means of matching the compliance of the actuation system with the static and dynamic duty cycle demands of the load. This can be achieved by interposing a mechanical transformer between the actuator and its load. An additional advantage is that the actuator can operate from a lower pressure source thereby allowing for maximum energy utilisation from a particular storage system.

Let us now consider a missile fin actuation system (3). Such a system operates for only a short duration and electro-pneumatics are an attractive option. A novel solution for such an application is shown in Figure 6. This shows the cross section of a unique combination of an orbiting motor and planocentric gearing which forms a particularly effective means of overcoming the limitations of pneumatics for a servo system. The output shaft is connected to the outer gear shown in the diagram. Within the outer gear is the orbiting cylinder block, its outside diameter being machined with a smaller number of teeth than are machined on the outer gear. The two gears mesh at just one point, this point moving around the outer gear in response to orbital cylinder block movement. This orbital movement is defined by the eccentric bearings on the locating crank arms. Pistons are situated within the five cylinders of the cylinder block and their hollow rods bear against a stationary hub located on the rotational axis of the output shaft. Differential control pressure is supplied to the hub and the application of a higher pressure to one half of the motor compared to the other half generates the force on the cylinder block. Correct phasing of pressures by means of the commutator valve allow this force to have a net effect which is normal to the radius of eccentricity. As a consequence, the cylinder block orbits about its radius of eccentricity resulting in the outer gear/output shaft rotating.

There are other virtues of this configuration beyond the obvious one that the orbital output of the actuator matches the orbital input requirements of the transformer. Firstly the effective motor inertia is mainly that due to the orbital mass about the radius of eccentricity rather than the rotational inertia of the cylinder block. The result is an inertia value which is typically two orders of magnitude lower than the value for a rotating motor of the same installation volume. Also,at the same speed of rotation, torque output is twenty to fifty per cent higher. The combination of these attributes along with the large mechanical advantage offered by planocentric gearing (typically twenty to thirty to one) results in a system with considerably higher dynamic response than more conventional electro pneumatic systems.

Disadvantages associated with this particular configuration are the higher cost because of the system's complexity, the need for a four way torque motor driven servo valve and of course the rotating commutator valve already described. These components also add to the possibilities for leakage. However, although the cost of such a system is higher than that of a conventional electro pneumatic system the advantages of high performance and relatively low cost compared with alternatives makes such a system very attractive.

5 DISCUSSION

Let us now try to compare the various merits of the three alternative types of control systems. To do so, we must compile a list of criteria that any user would consider when making his choice. A typical list would contain at least the following criteria:

- customer preference
- cost, size and weight
- environment (vibration, shock, acceleration, temperature, nuclear hardening,E.M.I.)
- others (storability, reliability, transportability, maintainabilty, re-usability and verifiability).

Let us now consider each of these criteria in turn. Customer preference or bias, is not an aspect for discussion in this paper. Cost is perhaps the most universally important criteria in the list. If we consider the fundamentals of a new servoactuation system and include all the elements that convert an electrical command to a controlled output, and also include in the equation the power source and power converter, then it can be shown that the following cost relationships generally apply. Taking an electrohydraulic system as the datum, a brushless samarium cobalt electromechanical system is likely to cost between eighty and ninety per cent of the datum price. A high performance orbital electropneumatic system such as described would represent a cost of approximately sixty per cent of the cost of the comparable electrohydraulic system. The lowest cost solution, a conventional electropneumatic system would probably represent fifty per cent of the datum. Clearly, if a particular solution is required which demands state of the art technology then cost differentials can be rapidly eroded and the foregoing statements no longer apply.

Size and weight comparisons are difficult to perform. In general terms the particular importance of one or other or indeed both of these criteria would not necessarily rule out any one particular solution. If however a system is being designed to be constructed from proprietary off the shelf components, then of course weight or size of a particular configuration may prove to be unacceptable.

Duty cycle considerations include static and dynamic performance characteristics, shelf life and operational life. If we first of all consider comparisons against the criteria of operational and shelf life, the virtue of each type of system is more or less apparent. However, quantifying the dynamic characteristics of different servo systems requires some clarification.

Figure 7. shows a variety of aerospace servo control requirements superimposed on a graph of control power against control bandwidth frequency. Based on our experience the technological limits of the three systems we have been considering are superimposed on the graph. The electrohydraulic option is available to all the various applications when one considers the dynamic performance requirements. It is really the only option open to a number of the applications. The threshold shown as the limit of electropneumatic technology relates to the sort of system described earlier. Any approach using more traditional technology would result in considerably reduced dynamic performance for a specific amount of control power. Similarly the threshold shown for electro- mechanical systems relates to the use of rare earth brushless technology. Any design using the more conventional brush type technology would have reduced dynamic capabilities.

Consideration of environmental and other criteria in the selection process could radically alter the choice of system for a multitude of reasons and consequently it is not proposed to add further comment to those made earlier.

Clearly the choice between electro-hydraulic, electromechanical, or electro-pneumatic control is not necessarily a simple one. If one is allowed to approach a problem with a completely open mind then the final choice may be made on the basis of what was initially considered to be a relatively trivial criteria. The importance of sophisticated electronics in supporting the various forms of technology discussed in the paper should not be underestimated. The ability to simplify the mechanical design of all the technological groups by the use of more advanced electronic design is a trend that will continue. Aerospace Fly by Wire systems are evidence of this trend and the sophistication now available in such systems will gradually be seen in more routine industrial applications.

ACKNOWLEDGEMENTS

The authors would like to acknowledge the help and advice they have received during the preparation of this paper from their colleagues at Moog Inc. in the USA, and at Moog Controls Ltd, Tewkesbury.

REFERENCES

(1) WAFFNER, W.B. and CHENOWETH, C.C.New Generation Flight Control Systems, Hydraulic Actuation. 96th Meeting of the SAE A-6 Aerospace Fluid Power and Control Technologies Committee. May 1984

© IMechE 1985 C268/85

(2) DAVIS, M.A. High Performance Electromechanical Servoactuation using Brushless D.C.Motors. Motor-Con '84 Conference, Atlantic City, New Jersey. April 1984

(3) THAYER, W.J. Electropneumatic Servoactuation - An alternative to Hydraulics for some low power applications. Fluid Power National Educational Seminar, Iowa State University. August 1984.

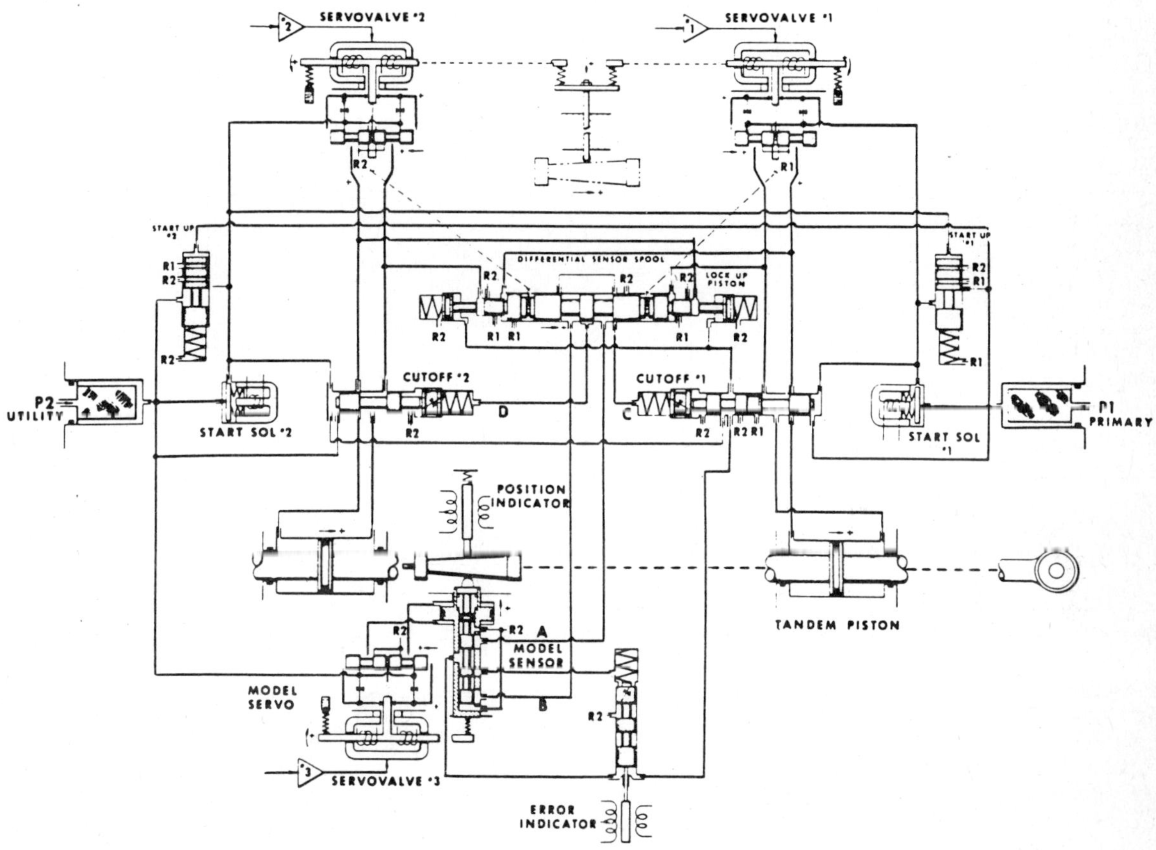

Fig 1 Conventional flight control system with redundancy

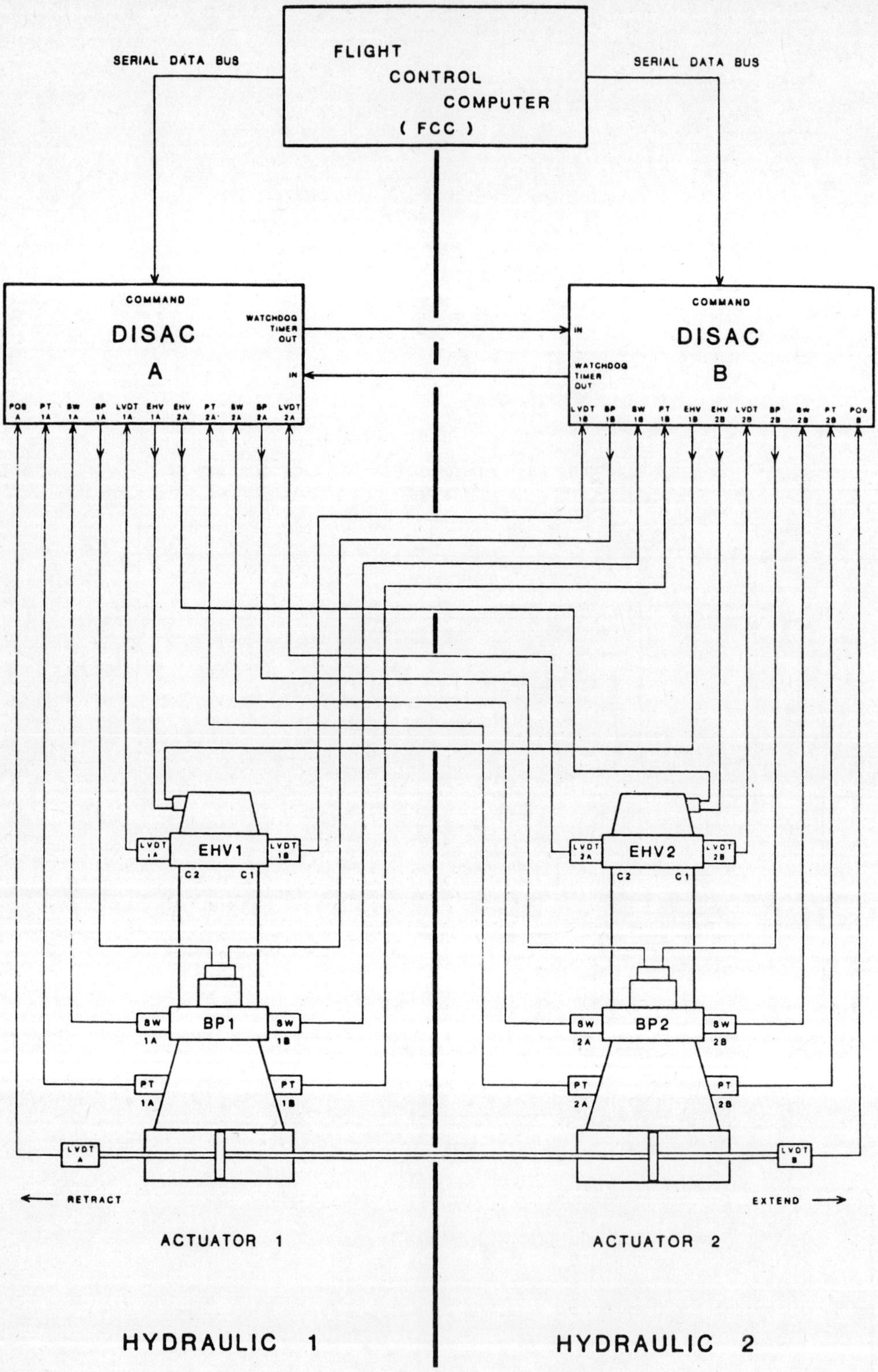

Fig 2 Microprocessor controlled hydraulic flight control system with redundancy

© IMechE 1985 C268/85

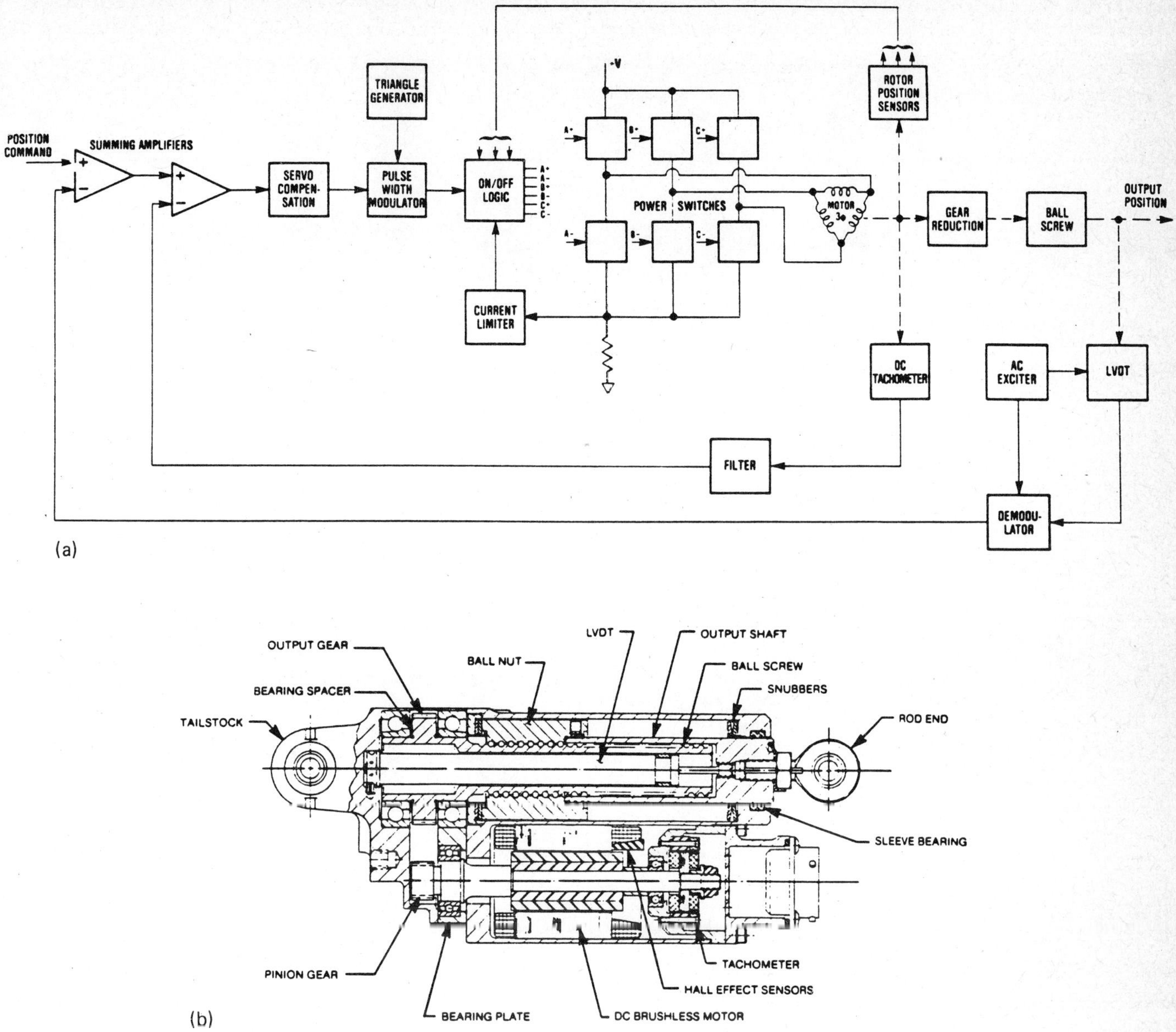

Fig 3 Electromechanical actuator (a) block diagram
(b) sectional drawing of actuator

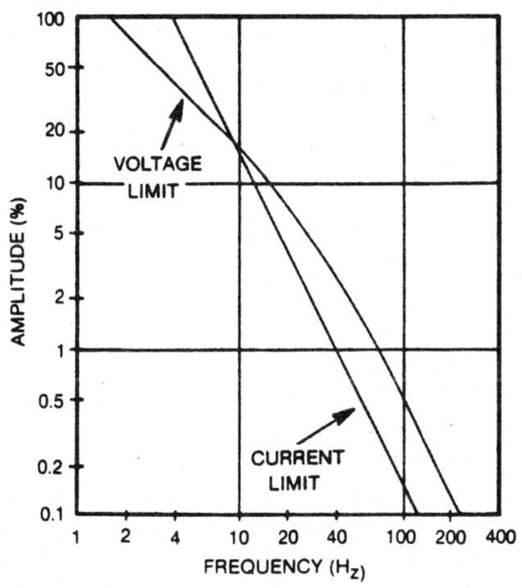

Fig 4 Response limits for electromechanical actuator

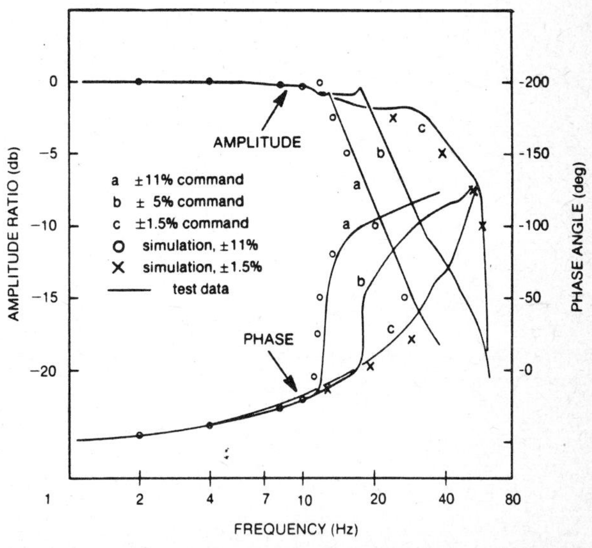

Fig 5 Response characteristics for electromechanical actuator

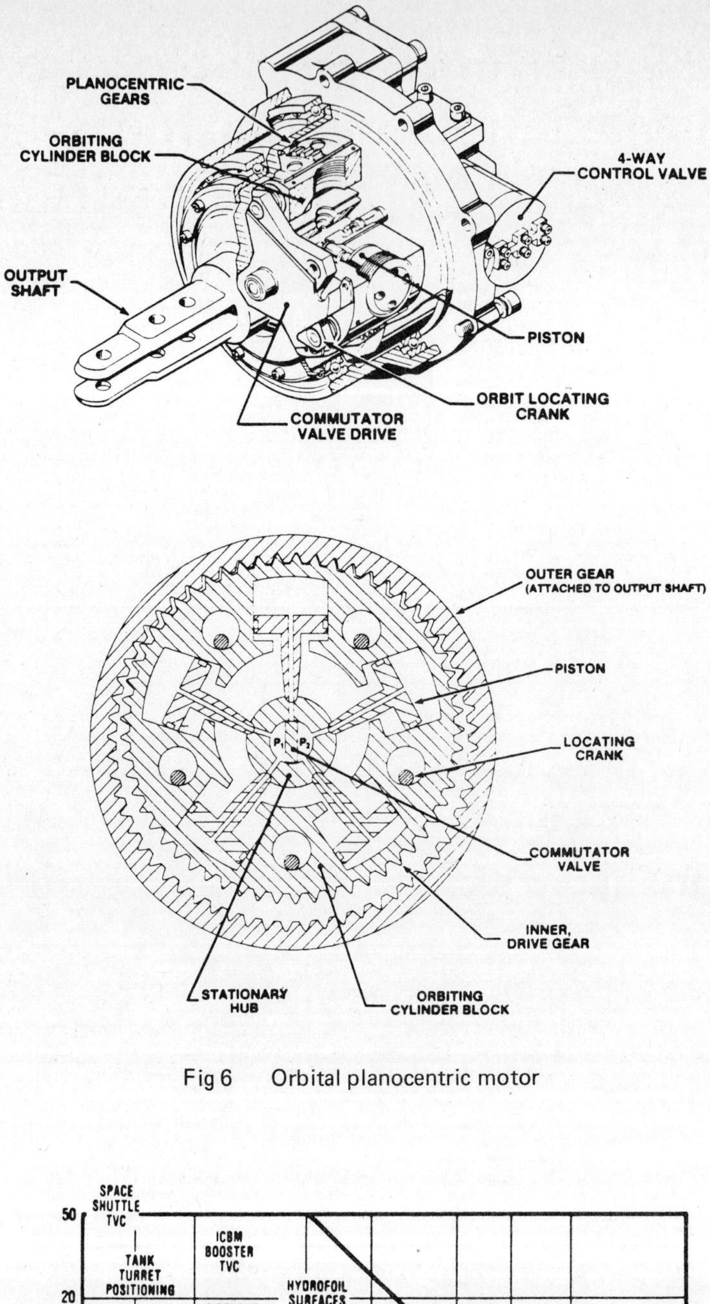

Fig 6 Orbital planocentric motor

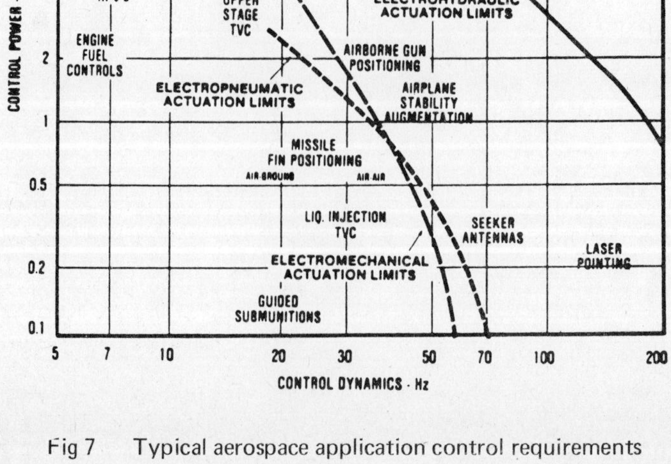

Fig 7 Typical aerospace application control requirements

© IMechE 1985 C268/85

C267/85

An electrically powered servo to drive the two axes of a missile launching platform – pros and cons when compared with the traditional hydraulic solution

J WALKER, BSc(Eng), **B P WARING**, BSc(Eng), CEng, MIMechE
and **M J CAHILL** BSc(Eng), CEng, MIMechE
British Aerospace Naval Weapons Division, Bristol

SYNOPSIS Recent developments in both power electronics and d.c. motor design suggests that the advantages of hydraulic over electric servo drives is narrowing. Alternative designs for a dual-axis missile launcher servo were investigated, using the following criteria: rapid traverse; accurate tracking; stiffness and stability; E.M.C; reliability and maintainability; total ownership costs; component availability; size/mass; cooling. The study showed that, to approach the weight and volume efficiency of the hydraulic design, the electric servo should use a P.W.M. amplifier driving an electronically-commutated rare-earth-magnet motor, and liquid cooling. Overall, the electric drive showed a disadvantage in terms of weight, volume and electrical interference, against better reliability, and through-life-costs.

1 INTRODUCTION

Our study is a comparison between Hydraulic and Electric Servos for a particular application. The application is relevant since the conclusions might be completely reversed if the application were changed.

Our application is the closed-loop servo-system which controls the azimuth and elevation angles of the missiles on a launching platform.

This launcher is part of the British Aerospace Rapier Guided Missile system. (see Figure 1).

To understand better this comparison a brief introduction to Rapier would be valuable.

It is designed to defend against low-level aircraft. It is capable of engaging helicopters which 'pop-up' and supersonic aircraft. Thus a rapid response system is essential.

The system is towed behind a Land Rover, manhandled and helicopter lifted. Minimum weight and size are important.

It has a deployment flexibility, frequently resulting in single system operation. Good maintainability and reliability must be achieved.

Since the system must provide its own power, efficient usage has to be made of the available power in order that generator size, fuel consumption and heat dissipation are minimised.

A large proportion of the Rapier System is a complex electronic system. Equipment used must not cause electrical interference problems. Also the equipment used must not be susceptible to interference in what is a very difficult RFI environment.

Rapier is designed to operate worldwide and so is capable of withstanding extremes of temperature and a wide variety of environmental conditions.

The system sells in a marketplace where international competition is fierce. Costs must be competitive in terms of initial purchase price and maintenance.

Rapier is a system with high performance yet is rugged, light and reliable. In technological terms it lies somewhere between an aeroplane and a tank, being somewhat closer to the former.

2 REASONS FOR MAKING THE COMPARISON

The existing Rapier Launcher servosystem uses a constant pressure hydraulic system.

The immediately obvious advantage of getting rid of hydraulic oil has often led us to consider the possibilities of using an electric servo. In the past we always concluded that an electric system could not replace the hydraulic unit without incurring unacceptable penalties.

However, successive considerations did reveal a narrowing of the gap. Improvements in power transistors and the availability of rare earth magnet electric motors were significant factors. Our customer, the Ministry of Defence to whom we are indebted for giving us the opportunity to carry out this comparison, showed considerable keeness to eliminate the hydraulic system being willing to pay penalties for what they believed would be an overall improvement, mainly obtained by simplified maintenance.

A comprehensive comparison was made in order that performance differences might be established and the total ownership costs for both systems determined.

3 PERFORMANCE REQUIREMENTS

The Launcher Servosystem has three distinct modes of operation which determine its performance requirements.

* Rapid response - it must slew a load onto an azimuth and elevation bearing rapidly, yet consume minimum power. (Shaft power from the engine must not exceed 7 KW).

* Stiff Launching Platform - when a missile is launched it applies shock loads to the servosystem. Launcher angular movements will cause the missile to deviate from its intended flightpath. Excessive deviation will result in an abortive firing.

* Smooth Tracking - when the missile has left, the launcher must smoothly track the missile in order that guidance data can be transmitted. In order that this smooth tracking is achieved we attempt to have gain margins better than 10 dB and phase margins better than 60°.

All the above must be achieved with the following complications:

(i) The inertia of the load varies depending on the number of missiles on the launcher. Maximum variation is by a factor of 10 to 1.

(ii) The out-of-balance load, which is considerable, also varies for the same reasons.

(iii) The large temperature range over which the equipment must operate results in characteristic changes.

(iv) The lightweight structure design results in structural resonances, whose frequency varies depending on the inertia.

(v) An interchangeability of equipment capability by simple plug-in, no setting-up.

4 SYSTEM DESCRIPTION

The mechanical arrangement to provide positioning in elevation and azimuth is as follows.

ELEVATION

The missiles are mounted in launch rails at either end of a cross-shaft. The shaft is located in bearings on a turntable and driven in elevation by a linear actuator, linked to a torque-arm.

AZIMUTH

The turntable is mounted on a slewing ring and is driven in azimuth by a rotary actuator driving a pinion meshed with the slewing ring.

A description of the servo systems is provided below.

4.1 Electric Servosystem

A schematic diagram of the electric servosystem considered is shown in Figure 2.

A brushless Samarium cobalt d.c. electric motor is chosen since it provides the lightest, most compact unit for our application. The inside-out configuration gives good cooling performance and the samarium cobalt enables high overloads to be applied for a short time, which suits our requirement for rapid slew. Liquid cooling of the motor is used, again to minimise size, but also to suit our installation. A step-down gearbox is employed to provide an overall ratio of 250 to 1 between motor and load.

The power amplifier used is of the pulse width modulated type which is supplied from a 290 V d.c. supply. Switching frequency used is approximately 4 kHz. The 290 V supply is achieved by direct rectification of the 200 V, 3ø, 400 Hz supply.

The servo pre-amplifier processes the signals digitally, using a dedicated microprocessor to provide the correct filtering and decision making.

Velocity and position feedbacks are derived from tachos and synchros (with digital conversion) mounted on the azimuth and elevation instrument gearboxes.

4.2 Hydraulic Servosystem

A similar arrangement is used for the hydraulic system. A constant pressure hydraulic supply replaces the 290 V d.c. supply. The pulse width modulated power amplifier is replaced by a servo valve.

The electric motor plus step-down gearbox is replaced by a slow-speed hydraulic motor which drives via a pinion directly onto the ring gear (overall gear ratio is only 11:1).

A hydraulic ram is used for the elevation drive.

A schematic of the hydraulic system is shown in Figure 3.

5 COMPARISON PARAMETERS

A hydraulic and electrical system were designed as described in section 4 and relevant data extracted in order that a comparison could be made. It should however be pointed out that any direct comparison of results is tempered by the fact that the two systems have not been derived from comparable data bases. Both equipments were designed with the aid of computer models. However, that for the hydraulic hardware was based on many years of service experience in diverse climates and considerable R and D, such that the tolerances due to environment and production scatter were well defined. In contrast, the model of the electrical system was supported by limited practical testing of a single system under laboratory conditions only. The test work carried out on this electrical equipment is covered in this section.

© IMechE 1985 C267/85

5.1 Servo performance

Both the hydraulic and the electric servos were designed to meet the performance requirements described in section 3. System hardware requirements were derived by performance computer models of both the electric and hydraulic systems. The hardware was subsequently designed and from it the information (e.g. mass and volume) derived. However the fundamental differences between the two systems complicate the issue.

A hydraulic system of the type described has an inherent stiffness (limited by leakage). The electric system must rely on its feedback gains to provide stiffness to external loads.

The servo must provide rate and position stiffness to meet our requirements. However we were unable to prove that these feedback gains were compatible with our structural resonances.

In order to maintain the power consumption of the electric servo within the figure specified, constant power control has to be incorporated during slew. Simple current limiting would cause excessive power demands. Our computer simulation showed that this could be achieved whilst maintaining a stable system. However, to our knowledge, this has not yet been achieved in a practical application.

Within the accuracy of our comparison we concluded that an electric servo is feasible to meet the performance requirements, as is the hydraulic system.

5.2 Mass and volume

Both systems were designed for minimum weight and volume for the reasons outlined earlier. On the electric drive the step-down gearbox and the motor were integrated into a single package. Also an isolating transformer is not employed due to its bulk. The weight comparison is shown in Table 1. The main weight differences can be accounted for as follows.

Azimuth Actuator - the electric unit must incorporate a gearbox, the hydraulic motor does not.

Elevation Actuator - the hydraulic unit is a simple hydraulic ram with servo valve, the electric unit is a recirculating ball screw actuator plus motor.

The PWM controller plus d.c. power supply is a heavier unit than the equivalent hydraulic power supply due in part to in-line filters which have to be incorporated to minimise conducted electrical interface. Filters which can carry the currents we are dealing with are large.

The volume penalty is mainly associated with the size of the PWM controller and power supply.

The electric system is approximately 1.8 times the weight and 8 times the volume of the hydraulic system.

5.3 Cooling

The Rapier system has to meet the specific duty cycle of operation. The most severe case is training in hot conditions (50°C). Under these conditions, continuous use of the servo is restricted to twenty minutes, of which five minutes would be slewing (high power) and fifteen minutes tracking (low power).

An intrinsic feature of a hydraulic system is that the hydraulic fluid also aids cooling, conducting heat away from the areas where work is done. Therefore individual actuators do not need external cooling, rather all the heat is taken back to the reservoir where it can be efficiently dissipated.

The version of Rapier considered incorporates a liquid cooling system for the electonics, and this has been utilised to also cool the servo. Figure 4 shows the relative effectiveness of cooling the electrical servo using either water/glycol as a cooling medium, or restricted natural convection. The term 'restricted natural convection' is used since it is not permitted to blow ambient air into the trailer structure, due to potential contamination problems. It may be seen that to maintain winding temperature at a safe level, liquid cooling is essential.

Liquid cooling is utilised for the hydraulic systems also, however in this case it is by no means essential as hydraulic fluid could be pumped to an external fan-cooled heat exchanger. However even when both servos are compared using water/glycol as an intermediate cooling medium, the electrical system has to pay a weight (and volume) penalty in providing the cooling circuit. The electrical system requires cooling fluid to be pumped to the two actuators and the electrical controller, whilst for the hydraulic system the heat transfer to the water glycol is achieved via a single cooling coil in the hydraulic reservoir. The additional mass associated with cooling the electrical system is illustrated in Table 1.

5.4 Electrical system integration

5.4.1 Power supply interaction

As said previously Rapier is small and lightweight and carries its own power supply which also has to be of limited size.

Also, in order that weight is minimised it is impractical to isolate the electric system from the supply by using an isolating transformer.

Therefore it is likely that our lightweight a.c. (400 Hz, 3ɸ, 200 V) supply will be affected by the demands of the electric servo power supply.

The power supply comprises a three phase full wave rectifier bridge and a reservoir capacitor. Once the capacitor is charged it will supply current continuously to the controller output switching devices. The capacitor will recharge when the input a.c. voltage exceeds the d.c. voltage across the capacitor. This will result in all the current to power the servos being drawn from the a.c. supply in short duration, high current pulses. The effect is illustrated in Figure 5. The inductance of the a.c. generator will tend to oppose the sudden changes in current and the result will be waveform distortion.

Measurements were made on a Rapier generator. Waveform distortion was increased from 1 per cent to 6 per cent when an electric servo load was placed on it.

This distortion will increase synchro errors and reduce the efficiency of the a.c. devices.

The hydraulic system has no adverse effect on harmonic distortion since the pump is driven directly from the generator engine.

5.4.2 Electromagnetic Compatibility

As the electronics used in weapon systems become more complex, it is important to ensure that problems are not created by mutual interference between equipment installed within a system and between systems used within relatively close proximity to each other.

The prospect of an amplifier switching nearly 300 volts at several kilohertz being installed in a compact weapon system was initially worrying and it was envisaged that EMC would be a considerable problem with the electric system.

In order to assess the magnitude of the problem a commercial single axis electric servo system was tested in the EMC test facility at British Aerospace, Dynamics Group, Bristol. Radiated emissions and conducted emissions on the power supply cables were measured. Some improvements were carried out on the system prior to the tests; these principally consisted of additional screening for the interconnecting cables and the installation of the controller in an aluminium case.

Typical results of the radiated emission tests plotted against the limits given in BSI specification BS3G100: part 4: section 2 are given in Figure 6. Although improvements would be needed to meet the specification, these are relatively modest and should be achieveable in a system specifically designed for installation in a weapon system.

Conducted emissions on the power supply cables exceeded normal limits by much more significant amounts (up to 70 dB at 2 MHz) and considerable filtering would be needed to reduce it to acceptable levels. Since the complete power consumption of the electric servo has to pass through these filters, they impose weight and volume penalties.

It may be concluded, therefore, that the problems of radiated and conducted emissions can be overcome but at a cost weight and volume penalty.

The hydraulic servo system does not cause electro-magnetic interference, nor is it susceptible to it.

5.5 Component availability

Equipment supplied for military applications is required to comply with certain standards in order that reliability will be maximised. One such requirement is that plastic encapsulated electronic devices are not normally used due to the risk of moisture ingress and subsequent malfunction.

Industrial applications of Brushless d.c. Drives use 300 V, 60A plastic encapsulated Darlington Pair modules which yield a compact solution. However, currently Ministry of Defence policy (Defence standard 69-61 part 1) limits the use of such devices to environments considerably more benign than they would experience in a Rapier launcher.

However, discrete power transistors of adequate rating are available from UK manufacturers with adequate approval for our needs (BS9000 and similar). These components are our choice for application in the electric servo. They are, however, somewhat bulkier when mounted than their Darlington Pair Power Module counter parts.

There is no difficulty with component supply for the hydraulic system.

It is concluded that component availability is not a serious problem with either system. There is, however, less practical experience of the power electronic devices.

5.6 Maintainability

The method by which a system is maintained affects the quantity and type of spares carried. Military equipment follows a maintenance plan where large sub-assemblies are interchanged in the battlefield without the need for significant adjustment. Ideally units should be of the plug-in nature with zero setting up required.

At the other extreme we have to return to the factory where detailed repairs down to component level are carried out.

In between there is a system of field workshop and in-country base workshop through which defective/repaired equipment passes.

The closer to the battlefield that repairs can be carried out to the lowest level the better since this will reduce the time to turnround defective equipment and consequently the number of spares which have to be carried.

Diagnosis of faulty equipment in the battlefield is generally carried out using test equipment built into the prime equipment.

General conclusions which might be drawn in this area are that hydraulic systems are not repaired easily in the battlefield due to the need to provide clean conditions before exposing a system. The electrical system lends itself more readily to the concept of interchanging units and to automatic fault diagnosis.

© IMechE 1985 C267/85

5.7 Reliability

Reliability directly affects availability and is therefore of prime importance in a weapon. The consequences of failure of a piece of military equipment extend far beyond the costs of repairing the failure. Reliability also directly affects the number of spares which must be made available.

Military equipment reliability is estimated from component counts and totalling individual failure rates for the complete system.

Table 2 shows a summary comparison of the reliability of both systems.

A significant factor in this comparison is that a hydraulic motor is ten times less reliable than an electric motor.

The reliability of the electronic equipment dominates the overall reliability of the electric system.

The overall comparison shows the electric system to be 1.58 times more reliable than the hydraulic system.

5.8 Total cost of ownership

The total ownership costs of a system depend on the following factors:

* Initial purchase price.

* Cost of spares, which is a function of the initial purchase price and the reliability and method of maintenance.

* Maintenance Equipment - the purchase price and the running costs of the maintenance facilities.

* Maintenance Manpower Costs - this depends on the method of maintenance and the reliability.

* Costs of consumables - oils, fuels to provide power.

* Design and development.

Our findings were:

1. The electric system has a higher initial purchase price due, largely, to the cost of the electronics.

2. Spares for the electric system are more expensive. This relates to the higher purchase price modified by the better reliability.

3. The hydraulic system maintenance equipment is more expensive. Specialised hydraulic facilities have to be supplied.

4. The hydraulic system is more expensive to maintain. Specially trained hydraulic technicians are required.

5. There is no significant difference in consumable costs.

6. There is no significant difference in development costs.

Due to lack of precise information on manpower costs and deployment we were unable to establish the exact cost comparison. However, the indications were that the electric system would have a lower total ownership cost.

6 SUMMARY

See overleaf.

7 CONCLUSIONS

The advantages for an electric servo system for this particular application could be summarised as:

* More reliable.

* Total ownership costs lower and easier to maintain.

* Quieter more efficient operation over the operating temperatures range.

The advantages for a hydraulic system are:

* Lighter.

* Lower volume which brings added weight advantage when housing structures are considered.

* No significant electro-magnetic compatibility problems.

* No significant electric systems interaction.

The conclusions of the comparison are that an electric system is feasible for the Rapier launcher servo. It holds, probably, an advantage in terms of total costs of ownership but has risks associated with the lack of hardware experience.

An electric system looks likely to replace the hydraulic system in future launcher systems.

6 SUMMARY TABLE OF COMPARISON

PARAMETER	ELECTRIC SERVO	HYDRAULIC SERVO	COMMENTS
Performance	Comparable		
Mass	111 kg	65 kg	
Volume	Controller = 50 LITRES	PUMP + RESERVOIR = 6 LITRES	Controller requires extra volume for liquid cooling
Cooling	Electric motors require liquid cooling	Cooling coil mounted in reservoir	
Electrical System Integration	Probably an E.M.C problem with electric servo		High power switching at several khz
Component Availability	'State of the art' high-power electronics	Technology improvements in circuit design rather than components	
Maintainability	Probably easier to maintain an Electrical System		Electric components easier to maintain closer to the "Front Line"
Reliability	Electric servo 1.6 times more reliable		
Total ownership costs	Inconclusive; indications are that the electrical solution shows advantage		Initial purchase of electrical system vs greater maintenance loss

© IMechE 1985 C267/85

TABLE 1 Mass comparison between electric and hydraulic servo.

ITEM	MASS (kg) ELECTRIC SERVO	MASS (kg) HYDRAULIC SERVO
AZIMUTH ACTUATOR	32	26
ELEVATION ACTUATOR	22	11
PUMP/RESERVOIR		16
P.W.M. CONTROLLER	30	
CABLE/HOSES	5	10
COOLING	22	2
GENERATOR	NOT SIGNIFICANT	
TOTAL	111	65

TABLE 2 Reliability comparison - mean time between failures

(m.t.b.f. of overall hydraulic system used as unit).

ELECTRIC		HYDRAULIC	
ELECTRONIC CONTROLLER AND POWER SUPPLY	2.25	4.70	HYDRAULIC PUMP, DRIVE AND RESERVOIR
ELEVATION DRIVE UNIT	17.28	3.98	ELEVATION DRIVE UNIT
AZIMUTH DRIVE UNIT	9.45	2.98	AZIMUTH DRIVE UNIT
SHRINK JACK	45.07	21.60	SHRINK JACK
ELECTRICAL DISTRIBUTION	70.59	6.47	HYDRAULIC DISTRIBUTION
TOTAL $\sum = \frac{1}{2.25} + \frac{1}{17.28} + \frac{1}{9.45} + \frac{1}{45.07} + \frac{1}{70.59}$	1.58	1.0	TOTAL $\sum = \frac{1}{4.7} + \frac{1}{3.98} + \frac{1}{2.98} + \frac{1}{21.60} + \frac{1}{6.47}$

Fig 1 The British Aerospace Rapier surface to air weapon system

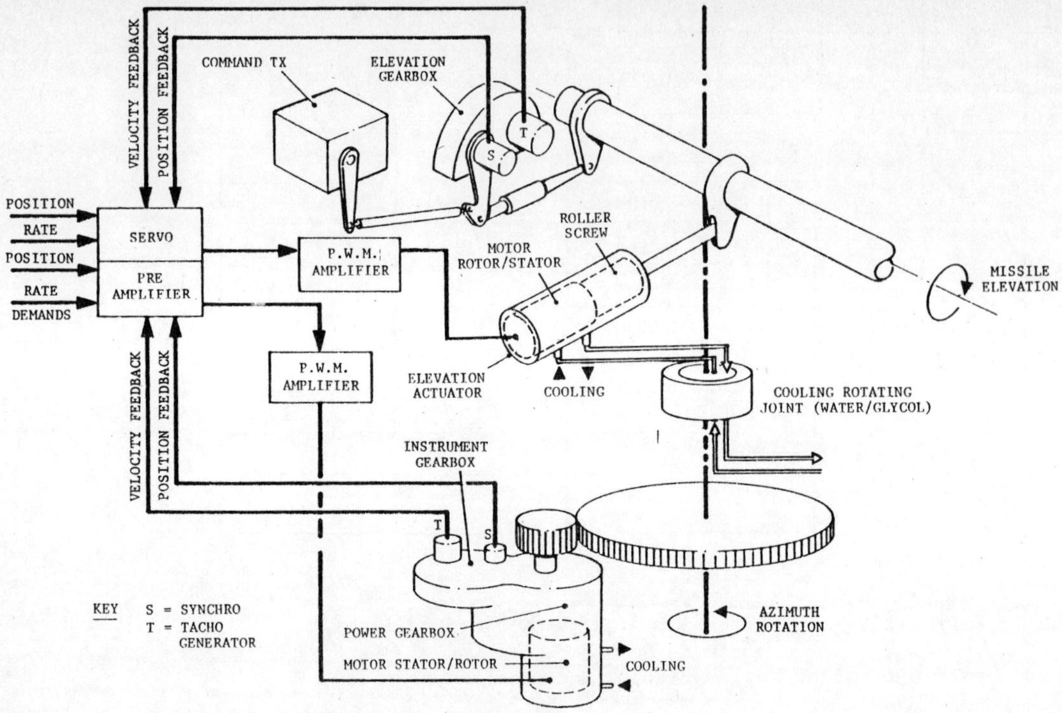

Fig 2 Overall system layout for the electric servo

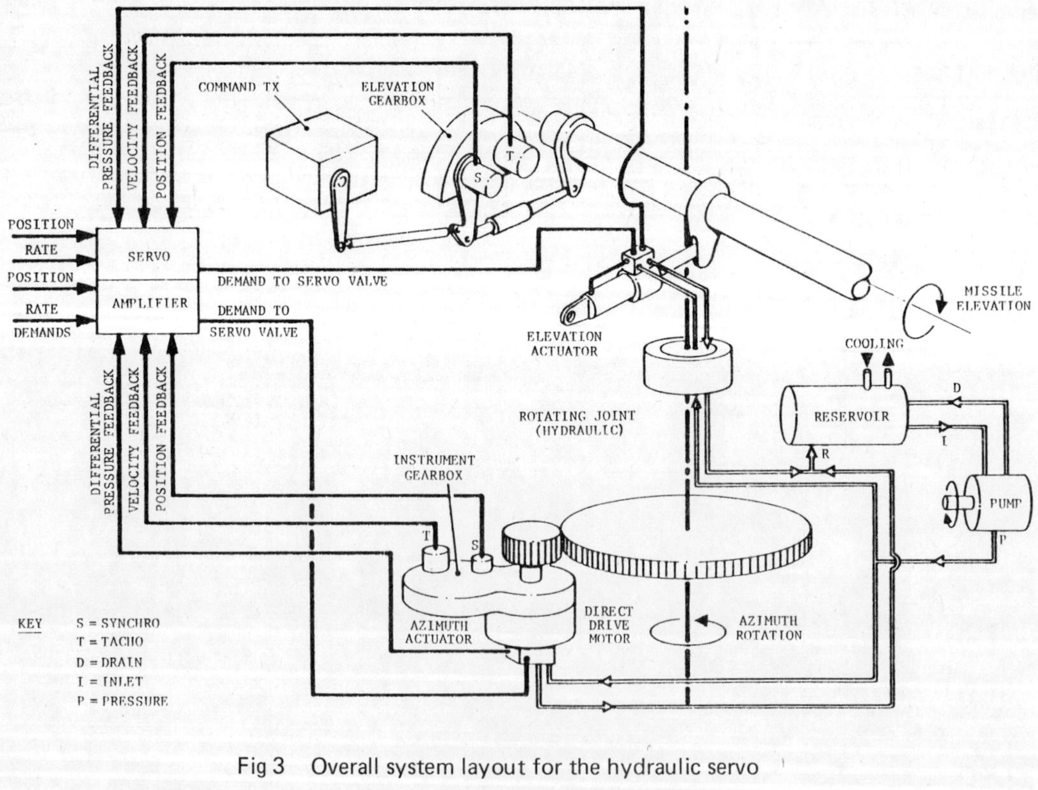

Fig 3 Overall system layout for the hydraulic servo

© IMechE 1985 C267/85

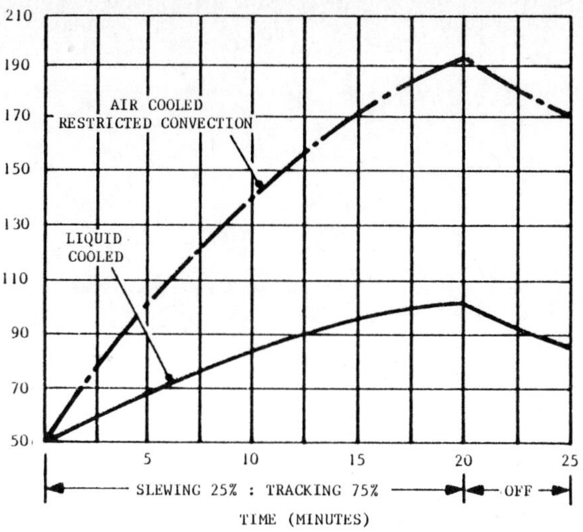

Fig 4 Thermal performance of the electric servo in a 50°C ambient – elevation axis

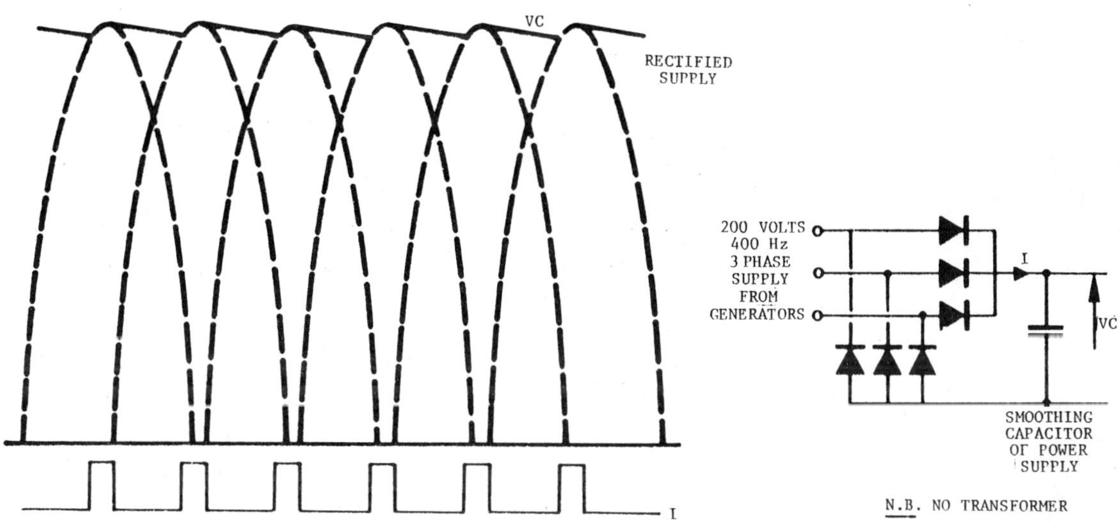

Fig 5 Power supply and current – idealized waveforms for zero-impedance source

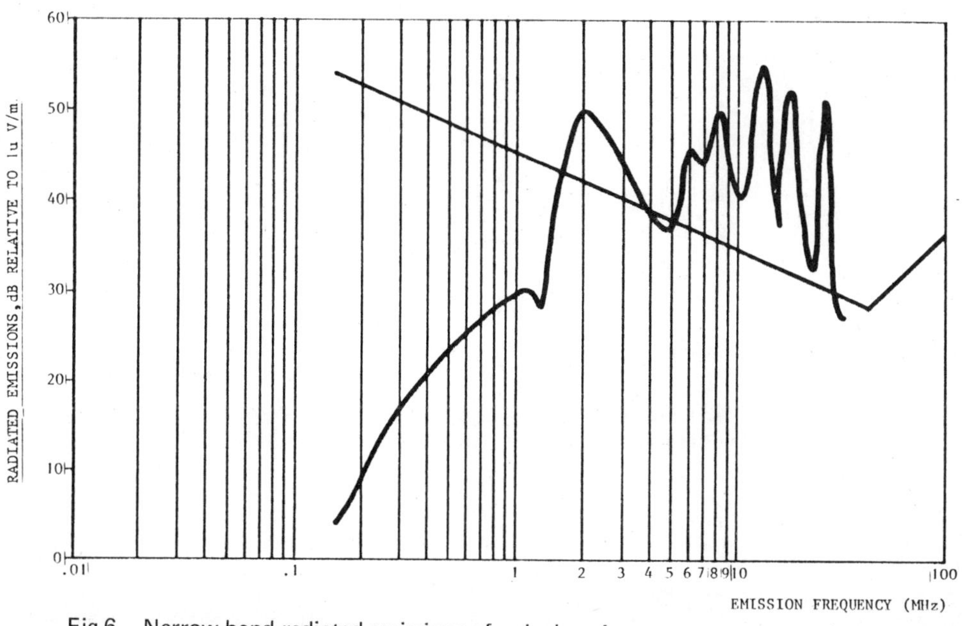

Fig 6 Narrow-band radiated emissions of a single-axis controller/motor

© IMechE 1985 C267/85

C270/85

The encapsulation of a novel intrinsically safe displacement transducer

SHEFFIELD UNIV.
APPLIED SCIENCE
LIBRARY

A RUSSELL,
Russell Associates
D P DRUDY, BA, TEng, MIED
Sacol Powerline Limited, Southampton, Hampshire

SYNOPSIS. Sacol Powerline Limited in conjunction with Russell Associates have developed a non contacting displacement transducer suitable for encapsulation within the piston rod of a hydraulic cylinder.

1 INTRODUCTION

Displacement measurement systems have been in use in servo hydraulic controls for many years. The system designer has the choice of installing the transducer either:

(a) externally - remote from the hydraulic cylinder, or

(b) internally - within the hydraulic cylinder piston rod envelope.

In the case of (a) difficulty has been experienced with backlash, vibration and differing geometrical displacements.

In the case of (b) long stroke displacement measuring devices have been either unavailable or exceedingly costly. Cylinder/servo actuator manufacturers have preferred to mount transducers coaxially as this has the added advantage of affording greater environmental protection.

FIG 1 shows a cylinder fitted with a coaxially mounted rectilinear potentiometer.

FIG 2 shows a Linear Variable Differential Transformer (LVDT) fitted in the tail rod of a hydraulic cylinder - remote from the pressurised fluid medium.

With the advent of the microprocessor and as lower cost electronic control systems developed, Sacol Powerline commenced a programme of investigation into the interfacing of electronics with hydraulic cylinders.

2 SPECIFICATION

(a) A measurement displacement device contained within the hydraulic pressure envelope, capable of signalling its position and velocity at all times.

(b) The measurement device should have a digital output of 0.1 inches resolution per bit, or alternatively an analogue output.

(c) The electronic interface must be able to communicate with microprocessors.

(d) To be compatible with mineral oil and high water based fluids.

(e) Its performance must not be affected by fluid borne contaminants.

(f) Have a low power consumption and be intrinsically safe.

(g) To be low cost in construction and readily available in any stroke length.

(h) Have good linearity, high accuracy and thermal stability.

(j) The device should not greatly increase the closed centres of a standard hydraulic cylinder.

(k) The device must be capable of transmitting signals over reasonably long distances.

A survey of displacement measuring transducers carried out in 1978 showed that current technology did not satisfy all of the specified criteria in one device.

3 EXISTING MEASUREMENT DISPLACEMENT DEVICES

The following list of advantages and disadvantages for each type of transducer has been compiled by the authors based on their experience in hydraulic applications.

3.1 Capacitance Transducer

Advantages. Simple modifications to cylinder, low cost.

Disadvantages. Difficulty in maintaining a consistant and reliable contact between moving piston and cylinder wall. Poor linearity and unacceptable end effects. Electronic tuning would result in many hours development work for each cylinder size. Variations in stroke requires modification to electronics. High water based fluids cannot be handled. Suspect stability with contaminated fluid.

3.2 Rectilinear Potentiometer

Advantages. Standard low cost units readily available.

Disadvantages. Special mountings none preferred. Long strokes difficult to obtain. Outside diameter increases with stroke. Suspect wear on contacts especially high frequency short strokes.

3.3 Linear Variable Differential Transformer (LVDT)

Advantages. High precision good linearity infinite resolution.

Disadvantages. Cost, limited stroke length, precision measurement depends on stability of electronics.

3.4 Magnetostrictive Transducer

Advantages. Large strokes up to 15 metres, simple installation, good resolution.

Disadvantages. Cost, long lead times for special strokes.

3.5 Encoders

Advantages. Standard relatively low cost products, good resolution.

Disadvantages. Requires mechanical lead screw attached to piston, difficulty in sealing rotating shaft at high pressures.

4 INITIAL DESIGN WORK

The initial work was based on the capacitance device. The limitations of such a device prevented further work ensuing and the alternative commercially available devices were in turn designed into the existing range of hydraulic actuators. Although these devices provided a short term solution, the long term aim was to design and develop our own measurement transducer which would satisfy the specification laid down in 1978.

4.1 IPHEX Survey

A survey was carried out at the 1980 IPHEX Fluid Power Exhibition which confirmed that in the higher technical industries, cylinders with coaxial transducers were well established. It was also evident that a low cost transducer would be well received by the industrial market.

The 1978 specification still stood with two exceptions. A digital device was now preferable and the resolution should be changed from 0.1 inches per bit to 0.5 mm per bit.

4.2 The Birth of The Linear Resolver

In January 1981 we were introduced to Alec Russell who had recently won an award for the use of a new plastic material called FLUOROSINT. This new material had been used as a basis for the construction of a new measurement device known as the LINEAR RESOLVER.

5 THEORY OF THE ROTATING RESOLVER

The concept for the linear resolver was derived from the rotating resolver FIG 3 which is well established as an angular measuring device. Two pairs of coils are fixed at right angles to each other and a drive coil is rotated within this field. If the drive coil is energised with a carrier current the device rotation will cause coupling in turn with:

(a) The first coil
(b) The second coil
(c) The first coil reversed
(d) The second coil reversed
(e) First coil again
(f) etc.

The electrical outputs at the two coils represents the carrier modulated by the sine/cosine coupling factor. If the carrier be considered to be Sin ω t the outputs can be considered to be:-

Sin ωt Sin θ and Sin ωt Cos θ

θ Being the mechanical coupling angle. Demodulation will provide Sine and Cosine signals which are conventional in data processing signals.

6 THEORY OF THE LINEAR RESOLVER

A linear helix was used to represent the drive coil in the linear resolver system shown in FIG 4. Reverse coupling was achieved by a double start winding which was shorted at the one end, so that both SIN ω t and. one half pitch away, SIN ωt + 180 degrees existed. Thus any search coil traversing the helix, double couples alternatively with these coils and produce SIN ωt SIN θ. A further search coil attached to the first in a spatial quadrature arrangement will produce SIN ωt COS θ thus simulating the rotary resolver.

The main difference between the natural rotary and the simulated linear resolver is one of coupling variations. The rotary action produces a pure sine/cosine coupling whereas the linear action is linear and coupling theoretically follows a triangular waveform law.

However, by careful selection of wire diameter and the spacing between the primary and secondary coils, and relative coil pitches, the triangular modulation can be modified into a sine wave form, which enables existing resolver processing electronics to be used.

7 LINEAR RESOLVER DEVELOPMENT

There were three hydraulic cylinder requirements which invalidated the original version of the linear resolver:-

(a) The resolver is essentially a two-part device (cylinder and piston) which must be mounted to both hydraulic cylinder and piston with cables attached. A cable can conveniently be attached to the hydraulic cylinder-end, but a cable connection to the piston is problematical in view of the hazardous environments in service. A single-cable version of the linear resolver was desirable (if not essential).

(b) Existing resolver was an air-cored device resulting in low-efficiency transformation and exceeding the drive current constraints for I.S. approval. Means to increase efficiency required.

(c) The varity of hydraulic cylinder sizes and strokes made some form of rationalisation of transducer parts necessary to cover the range.

© IMechE 1985 C270/85

7.1 Hardware Evolution

The requirement of (a) was satisfied by adding an energising or drive coil (3) see FIG 5 to the search or detector coil assembly and utilising a single-start measurement coil with shorted ends to encourage circulation of carrier current. Thus there is double transformation, the moving energising coil carries SINωt in a non-spatial manner which is induced into the shorted helix (1) and from there induced into the search coils (2) in a now spatial format. As the measurement coil (1) is now essentially single-start the search coils (2) require to be double-start helices. No cable is required to the shorted-coil section of the transducer.

The requirement of (b) was tackled in several ways, methods of screening the measurement coil to induce more current from the drive coil certainly increased the load on the drive coil but resultant eddy currents losses worsened matters. The best improvements resulted from the use of ferrite cores in the non-spatial drive coil when effective operation was observed with drive currents less than 30ma.

The requirement of (c) posed difficult production and operation problems. The production range was broken into two classes, one class could tolerate a hole of 20mm being bored into the piston and the other class a hole of only 16mm.

7.2 Description of Operation
1st Prototypes.
(See FIG 5)

The measurement coil (1) of known pitch is energised by induction from a suitable a.c. signal (4a) applied to coil (3). The detector coil (2) has two coils wound in spatial quadrature connected electrically at (4b).

As the measurement coil (1) moves relative to the fixed male detector coil (2), signals are passed to (4b) which represent the sine and cosine of the linear angle between the measurement and detector coils. The signals are then processed to provide a 0.5mm resolution per bit.

The device shown in FIG 5 has limitations because the length of the measurement coil is directly related to the length of the hydraulic cylinder stroke. On long stroke cylinders this coil would have a high inductance and require considerable power to circulate the necessary current.

7.3 Low Power-Breakthrough

A means was sought which would permit the linear resolver to operate at low power irrespective of length of stroke. The objective was satisfied in a transducer the embodiment of which is shown in FIG 6 in cross sectional diagrammatic form.

The measurement coil (5) now consists of a number of adjoining segments (five in the example) which are each short circuited on themselves electrically so that current can circulate in isolation. Power is provided by energising drive coils (6) which are situated on each side of the detector coil (2). The coils (6) are closewound, non spatial in behaviour and serve to commutate the measurement coil. Power at the chosen frequency is fed to coils (6) which are inductively coupled to measurement coil segments (5) in which current circulates. Never more than three coils are coupled so that very little power is required, irrespective of stroke length. This segmented construction enables intrinsically safe requirements to be met.

The search or detector coil (2) is spatially wound in quadrature and senses the spatial excursions of the measurement coil which is attached to the moving piston rod. The detected sine and cosine signals are available for processing at (4).

7.4 Materials

The outer measurement coil is contained within a steel tube which is fitted with a threaded ferrule at one end. This assembly is mounted inside the hollow piston rod.

Good thermal stability was the main criteria for the measurement coil and this was achieved by using a glass based composite. The material has excellent electrical insulation properties and low water absorption values.

The detector probe is manufactured from one of the many plastics widely available which offers good thermal stability and electrical properties. The probe is supported on a non magnetic stainless steel tube.

7.5 Specification-Change

By late 1983 the specification had been revised again. The resolution required had changed from 0.5mm per bit to 0.1mm per bit. This requirement coincided with the development of a CMOS processor eurocard, and was possible due to the quality methods of manufacturing the transducer which provided clean signals for subsequent processing. Subdivision of the detected signals was now only a matter of electronic wizardry.

8 DATUM RESET

The transducer is incremental and non absolute. It is however absolute over each pitch of the measurement coil.

A clever piece of design work has enabled a datum reset signal to be obtained from the standard transducer at one end of stroke. The datum reset signal is effective at the end of stroke. The feature operates by observing the centre tap of the drive voltage to the drive coils. On reaching the last 16mm of stroke one of the drive coils commences to disengage its coupling from the measurement coil, its loading reduces and causes the centre tap voltage to swing positive or negative. Thus a positive going ramp is generated at the cylinder extended end and a negative going ramp is generated at the cylinder retracted end. The ramps are 16mm long and after amplification are brought out for customer use. The voltage analogue reference Var is also brought out for ramp detection, this voltage will be equal to the ramp voltage when end of stroke ramp is absent (normal) operation.

A third output comprises a pulse which is coincident with the measurement coil pitch structure. This is very precise and can be gated with the ramp signal to produce an accurate datum signal.

A mid stroke datum has recently been developed which enables counters to be reset at preselected parts of the stroke.

9 ELECTRONIC DRIVE AND OUTPUT CARD

The electronic signal processing is contained on a standard 100 x 160 Eurocard fitted with DIN 41612 edge connector. An intrinsically safe version of this card is also available.

9.1 Inputs

The inputs are:-
regulated 12 -15v DC
unregulated 17.5-30v DC
supply current 70mA

These inputs are used to drive the transducer at 19 kHz with a primary energising voltage of 2.1v r.m.s. at 26mA.

9.2 Outputs

The processor provides BI-DIRECTIONAL serial outputs giving a measurement resolution of 0.1mm. Hysteresis equivalent to +/- 0.5 counts is built into the processor to prevent the occurrence of spurious (jitter) pulses on both outputs at critical rest positions. This may arise due to electrical background noise which fractionally changes the switching point of the analogue to digital interface circuit. While this would not lead to cumulative errors, the jitter effect is undesirable where a visual display of the position count is required.

Outputs are provided to give the user the choice of both commonly used modes of bi-directional counting:-

Mode 1: separate serial outputs, one for incrementing UP CLOCK, and one for incrementing DOWN CLOCK. The output pulses are non-overlapping and are active low in common with most counter input information.

Mode 2: a common bi-directional serial output U/D CLOCK and a separate directional control line.

For the standard processor 2 metres/second maximum pulse rate, the pulse duration of all serial outputs is 5msec or 0.25 x output pulse rate whichever is the smaller.

A complete list of Input and Outputs is given in Appendix A.

10 THE COMPLETION OF THE INTELLIGENT CYLINDER

The transducer has been designed into the 250 bar HYACS range of hydraulic cylinders, a typical cross section is shown in FIG 7. The cylinders are modular in construction and are also available with mechanical locks and stepper motor driven servo valves.

To complete the intelligence, a range of electronic controls are available, compatible with microprocessor control.

11 APPLICATIONS

The transducer is ideally suited to both open and closed loop systems.

(1) The output pulse train at speeds up to 20 KHz can be fed to a set point controller to achieve multiple switch points for logic control systems.

(2) Visual indication of cylinder position is readily available when used with a Counter.

(3) The resolution of 0.1mm makes it ideal for closed loop control using proportional valves for low cost positioning systems.

(4) The problem encountered with limit switches in dust laden environments are overcome. This has been achieved with a SACOL INTELLIGENT cylinder in the production of viscous fibre from wood pulp.

(5) Teach and learn systems for two or three axes control can be achieved at medium cost. One such system has already been developed and is undergoing trials.

(6) Aligning chock advance cylinders on a coal face.

(7) Monitoring roof movement on roof supports.

(8) Industrial Automation systems where logic sequential and positioning operations are required with a greater degree of flexibility.

12 CONCLUSIONS

The majority of objectives detailed in the 1978 specification have been met. The resolution is 25 times smaller than originally specified. In addition to mineral oils, the transducer will work in 100% water, and phosphate ester fluids. Filtration levels have been set to 25 micron in keeping with good hydraulic practice. The transducer and its processor meets intrinsically safe requirements. The basic length of a cylinder has increased by 40mm on the smallest 50mm bore size and by 50mm on the largest 250mm diameter cylinder. A cable length of 50 metres can be used between transducer and processor.

We are pleased to report that development work is continuing into finer resolutions. The electronics will soon be available on a 40mm x 40mm square thick film hybrid which will be encapsulated on the end of the detector probe.

12.1 Summary of Transducer Advantages

The 'System 100 micron' transducer has many advantages for cylinder operation, the combination of which in a single transducer is thought to be unique.

© IMechE 1985 C270/85

(1) Digital structure of measurement coil for permanent establishment of accuracy.

(2) Analogue signals for fine subdivision of digital structure.

(3) Operation in all hydraulic mediums including water-based fluids.

(4) Very low electrical currents.

(5) Measurement tube self-energised and requiring no electrical cables. (Single cable system).

(6) Completely non-contacting for long life cycle.

(7) Internally fitted for environmental protection.

(8) Associated CMOS electronics eurocard provides displacement information in digital form suitable for the majority of counting and control applications.

(9) Optional end-of-stroke or part-stroke feature for setting very accurate datum positions.

(10) Low cost solution for long strokes.

HYACS SYSTEM 100 MICRON TRANSDUCER ELECTRONICS

APPENDIX A.

PIN

2a - Supply oV

2c - TRANSDUCER SIN o/p Hi

4a - " SIN o/p screen

4c - " SIN o/p Lo

6a - " COS o/p screen

6c - " COS o/p Lo

8a - " PRIMARY DRIVE Lo (screen)

8c - " COS o/p Hi

10a - " PRIMARY CENTRE TAP

10c - " PRIMARY DRIVE Hi

12a - END-OF-STROKE Veos

12c - ANALOGUE REFERENCE Var

14a - NC

14c - NC

16a - EXTERNAL CLAMP VOLTAGE INPUT

16c - SUPPLY OUT (for V CLAMP =Vs)

18a - $\overline{\text{UP CLOCK}}$

18c - DIRECTION

20a - $\overline{\text{U/D CLOCK}}$

20c - $\overline{\text{DOWN CLOCK}}$

22a - WINDING REFERENCE

22c - $\overline{\text{DIRECTION}}$

24a - NC

24c - OV (LOGIC)

26a - N.C.

26c - N.C.

28a - N.C.

28c - N.C.

30a - UNREGULATED SUPPLY (17.5 - 30 V)

30c - N.C.

32a - INTERNAL REGULATOR OUT (+15V d.c.)

32c - REGULATED SUPPLY IN (15V d.c.)

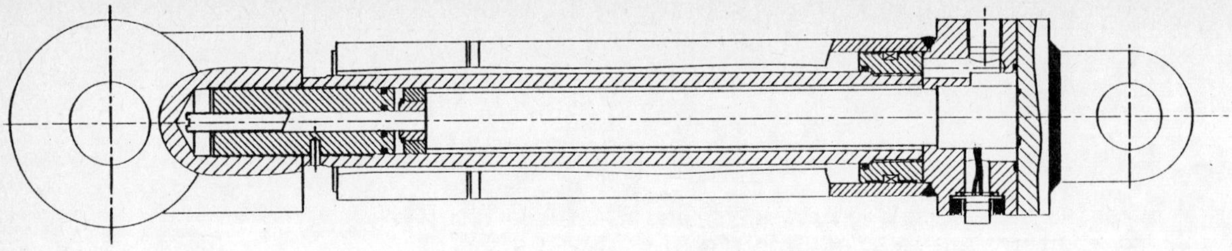

Fig 1 Hydraulic cylinder with rectilinear potentiometer

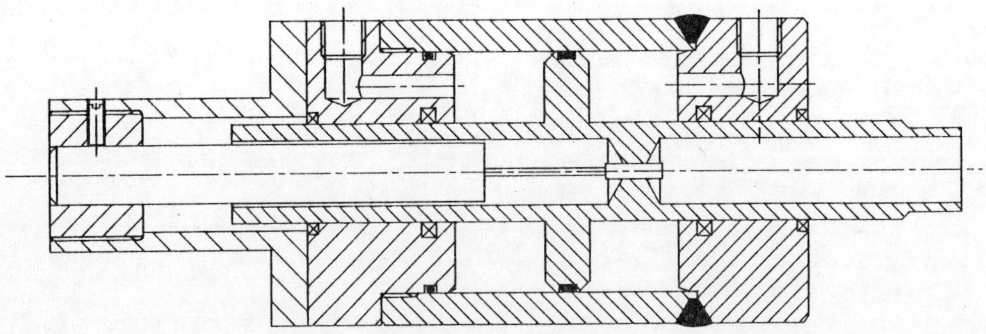

Fig 2 Through rod hydraulic cylinder with linear varible differential transformer

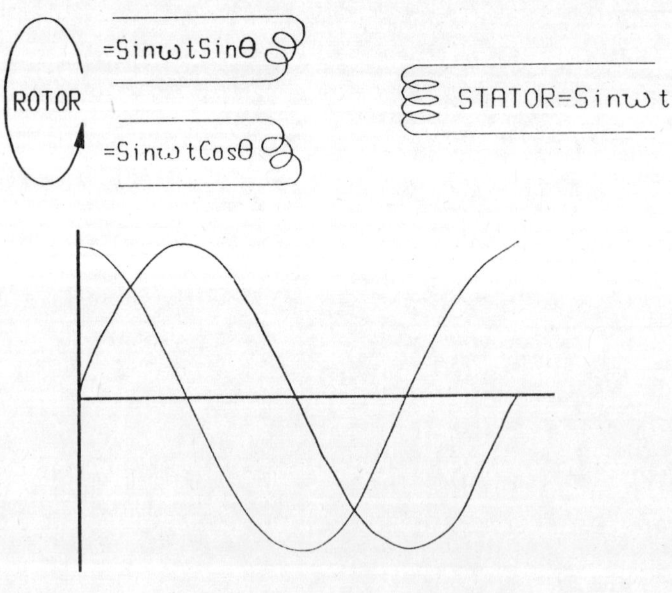

Fig 3 Rotary resolver with sine and cosine outputs

© IMechE 1985 C270/85

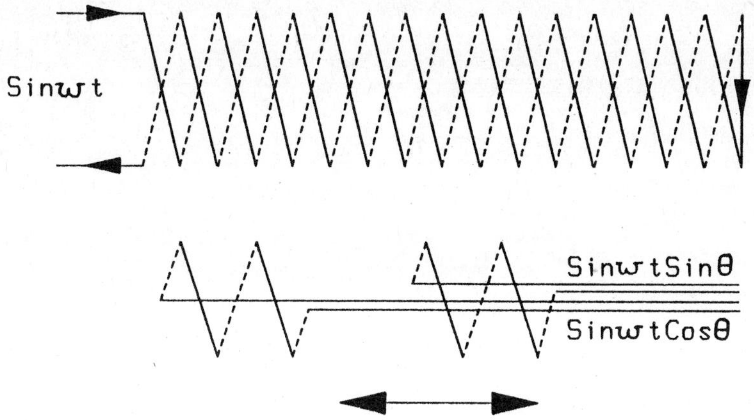

Fig 4 Linear resolver windings

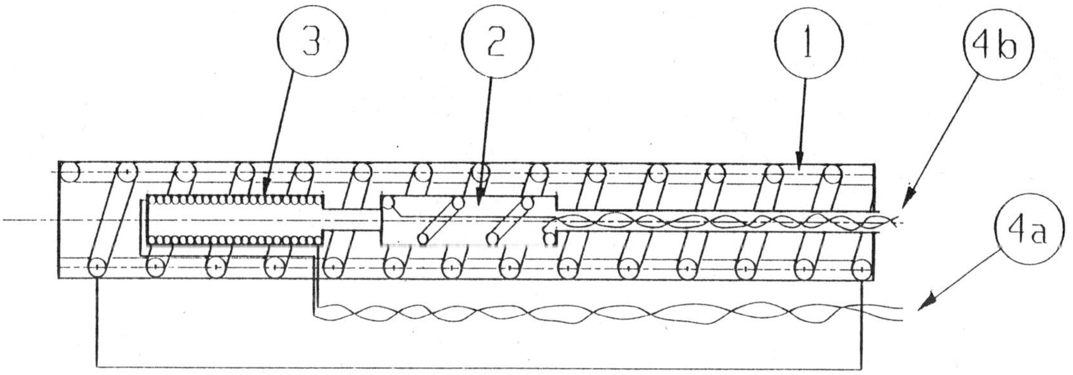

Fig 5 Early prototype linear resolver

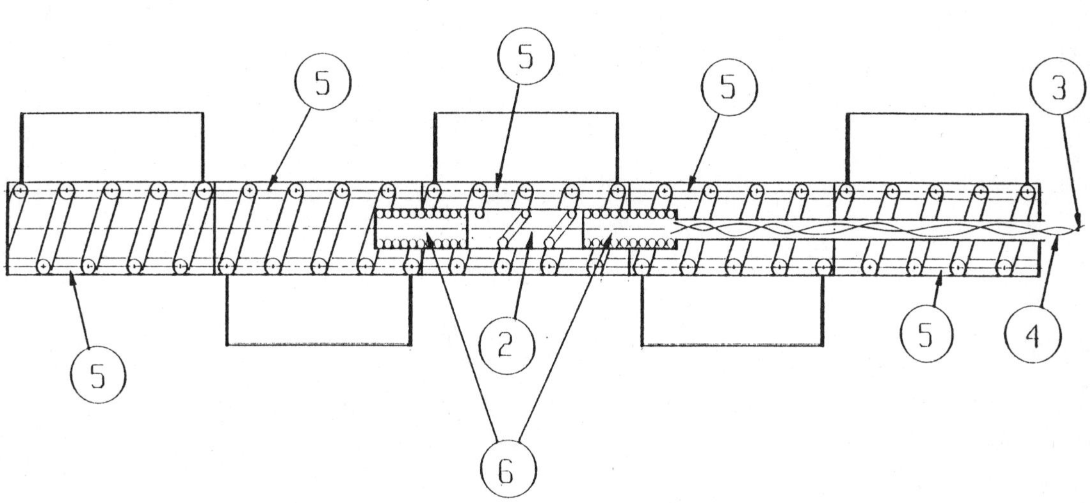

Fig 6 Low power version linear resolver

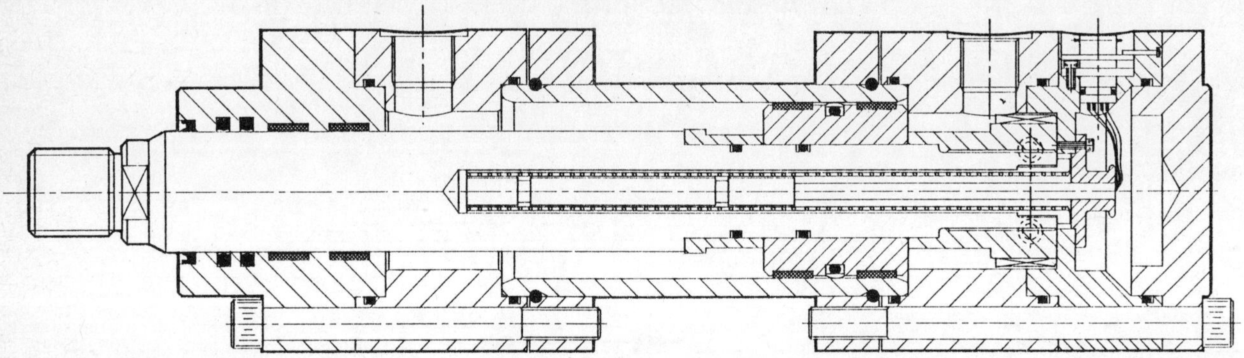

Fig 7 100 μm transducer fitted to Hyacs hydraulic cylinder

© IMechE 1985 C270/85

C273/85

Mobile cryogenic pumping systems

F FARRAR, CEng, MIMechE

Walter Jones and Company Limited, Sydenham, London SE

SYNOPSIS Whenever there is a requirement for a system to transmit power from source to point of application, there is, and always will be, a place for hydraulics, electrics and pneumatics, each bringing to industry its own special advantages, and of course its own limitations. The purpose of this paper is to examine in detail the application of hydraulic and electric systems of power transmission in a mobile cryogenic pumping operation.

MOBILE CRYOGENIC PUMPING SYSTEMS

It will probably be of some value to give a brief description of the products to be pumped that form the basis of this paper. A cryogenic liquid is normally accepted as any liquid with a boiling point below -160'C (113.15 K) at a pressure of 101 325 Pa (one standard atmosphere). Obviously there are extremes to this approximate classification. At one there is liquid natural gas with a boiling point of -161.61'C (111.54 K) which is literally shipped in tens of thousands of tonnes across the world. At the other extreme is liquid helium with a boiling point of -269'C (4.15 K) with only small quantities moved, measured in litres rather than tonnes.

For the purposes of this paper I propose to limit my observations to the movement of the three main liquid atmospherical gases used in industry today. They are liquid oxygen, nitrogen and argon. Liquid oxygen with a boiling point of -182.96'C (90.19 K) is, in itself, non-flammable, but of course it supports combustion even to the extent that stainless steel will burn in an oxygen enriched atmosphere and should an ignition occur in an oxygen enriched atmosphere, it is very difficult to extinguish. Liquid nitrogen at a boiling point of -195.80'C (77.35 K) is inert and does not support combustion or life. It is often used as an instant source of low temperature as in freeze grinding, freeze drying etc. Liquid argon with a boiling point of -185.86'C (87.29 K) is also inert and is mainly used in inert gas electric welding.

Distribution of these liquids from production plant to customer is usually accomplished by means of vacuum insulated road tankers with liquid capacities varying from 5000 litres to 28 000 litres, often fitted with a product pumping system to transfer the liquid into the customer's storage vessel. A pump has been defined as a machine which transfers mechanical energy from some external source to the liquid flowing through that machine. It is that transmission of power in these cryogenic pumping systems that we are now going to consider. Fig 1 shows a typical 24 000 litre cryogenic road tanker fitted with a pumping system suitable for the conveyance of liquid atmospherical gases.

Currently there are four alternative systems with varying degrees of popularity in service that are used to provide sufficient power to successfully operate the cryogenic pump, namely:

(1) Independent i.c. or c.i. engine

(2) External electrical power

(3) Hydraulic power using vehicle engines as prime mover

(4) Electric power using vehicle engine as prime mover

Because of the theme of this seminar it is appropriate that we compare in detail systems three and four, the hydraulic and electric. But first a few words about one and two. The independent engine fitted to the side of the tanker, with its own independent fuel supply, is directly coupled through a lay-shaft and gearbox to the cryogenic pump. This system has many disadvantages and is not widely used. The external power taken from the customer's electrical supply usually 50 or 60/hz, connected by a flying lead to an electric motor direct coupled through a gearbox to the centrifugal pump. This system has a number of advantages, simple to install, easy to operate, quiet. It also has many disadvantages and is only in limited use internationally, although it is fairly widely used in Europe.

The purpose of this paper is not only to report on the two main systems currently in use, electrical and hydraulic, but also to demonstrate their differences. Should a company contemplate a change in systems or a newcomer into the field seek guidance, the information contained here may be of value to enable the user to make a final choice more suited to the skills and facilities available. I have chosen the application of mobile cryogenic pumping systems because it is one of the few where a direct comparison can be made between the electrical and hydraulic method of converting and transmitting energy.

Throughout the world there are a number of cryogenic liquid producers supplying a greater number of customers, surprisingly the power requirement in each case is very similar, a UK tanker could quite easily serve say a Japanese customer. On an international scale there are very few manufacturers of cryogenic liquid pumps. Their customers, also being international, with dissimilar legislation for the conveyance of dangerous goods, contrasting priorities and safety principals, and a very different degree of available skilled personnel, requires a flexible approach by the cryogenic pump manufacturers when submitting quotations. In other words, almost invariably the manufacturers have the capability of supplying cryogenic pumping systems compatible with any of the four systems of power supply mentioned earlier, according to the customer requirements. The customers choice is therefore not limited by the availability of systems from the equipment manufacturer, but only by his own persuasion.

In the cryogenic liquid distribution industry the influences normally considered when purchasing or supplying equipment, cost, reliability etc. have different priorities, except of course where safety is concerned, which must always be of first consideration. The total cost of a vacuum insulated cryogenic road tanker can be as high as £150 000 with the ancillary equipment, pumps, valves, etc., only taking possibly 10 per cent of that figure. Therefore reliability in the ancillary equipment is far more important than the initial cost of the equipment.

The cryogenic liquid transfer rate from tanker to customer storage is usually in the region of 500 litres per minute (110 igpm) at pressure heads up to 350 metres. This is normally achieved by a single stage centrifugal pump rated at a speed of 7000 to 8000 r/min, absorbing approximately 45 kW. In certain circumstances a multi-stage cryogenic pump rated at 3500 to 4000 r/min is used, but this is obsolescent and is not now customary practice. Fig 2 shows a single stage centrifugal pump and Fig 3 a multi-stage submerged pump with hydraulic drive.

Let us now consider the various influences that must be analysed before adopting one or other of the systems available for a particular duty.

(1) Safety

(2) Reliability

(3) Efficiency

(4) Maintenance

(5) Total weight

(6) Operation

(7) Cost

(8) Availability

(9) Robustness

(10) Compactness

(11) Standard design

(12) Adaptability

All these points must clearly have a bearing on the final decision, but due to limitation of time I will confine my comments to what, in my view, are the four most important considerations; safety, reliability, efficiency and maintenance. I will of course be happy to discuss the remainder of the list at the end of the paper, should you wish to ask questions and time permitting.

(1) SAFETY

Whatever the process, whatever the industry, 'safety' is the first consideration through every stage of the design and planning required to bring a plant or process 'on stream'. In the specific application we are now considering however, personal safety has a very different connotation from that of personnel in say a factory, or fully automated static processing plant in a constant temperature, weatherproofed environment. The tanker driver/operator is generally semi-skilled, with probably no previous knowledge or training of chemical or process plant operation. Almost invariably he performs his duties without any kind of personal supervision, with no one on hand with specialist skills or product knowledge to personally advise him should a malfunction occur in the equipment. By any standards, the operation he performs is labour intensive and very dependent on his personal skills. A great deal of time and money is spent in trying to reduce the dependency of the equipment on the frailities of human nature, and considerable progress has been made, but until we have a Star Wars R2 D2 and C3 PO, the driver/operator is on his own and indispensible. It is then in this climate, that the decision to use the electric or hydraulic system of transmitting power from the vehicle engine to the cryogenic pump has to be considered.

Of all the systems mentioned, there is little doubt that the self-generated electrical system is the most preferred system. The power is generated by the vehicle engine, and transmitted through a power take-off into a gearbox integral with the electrical alternator. The alternator terminal box contains, overspeed, over temperature, and excess power, safety trips, with a small separate box containing isolator switch etc. The electric power is then routed from the alternator through an armoured cable to the control adjacent to the electric drive motor and centrifugal pump at the back of the tanker.

The operation of the equipment is controlled through a 24V circuit, with a simple push button on/off switch. When the driver/operator has completed the preparation of the tanker and the receiver by connecting the product liquid hose to the customer storage and engaging the power take off, he simply presses the start button, then concentrates on bringing the product pump on prime. He is confident that should a malfunction occur in

© IMechE 1985 C273/85

any part of the equipment power will be cut off safely and immediately.

Preventive maintenance checks on electrical equipment are non-destructive, electrical leakage for instance is continually monitored during operation, and should an electrical leakage exceeding 20mA occur, the electrical generating equipment will instantaneously and automatically shut down, and cannot be restarted until the fault has been cleared.

The concept of the hydraulic system is very similar to that of the electric system, but of course the power transmission is accomplished by a very different principal. Similarly to the electrical system, the vehicle engine is connected to a hydraulic pump by means of a lay-shaft through the engine power take-off. The pump transfers fluid from a reservoir through hoses and pipes to a control panel of valves, actuators etc., to a hydraulic motor, and finally probably through a gearbox, to the cryogenic pump. Fig 4 shows a simple hydraulic system.

Preparing the tanker and the customer storage for the transfer of liquid and the start of the pumping operation is very similar both for the hydraulic and electric systems. With the hydraulic system, although it is a gradual and smoother start, there is the knowledge that fluid leakage can occur at any time. Every piece of moving mechanical equipment has a characteristic noise, and in my view a difference in that noise is a clear indication that a change in condition has occurred requiring investigation, and which may well be a warning of impending failure of a component. In my experience a hose or coupling failure gives no such audible warning. Yes, of course there may well be a visual indication, bulging hoses etc., but as most of the equipment is out of view of the operator, a loud report and loss of power are probably the first indications of failure. Although pressure loss shut-down devices are normally fitted, I am not aware of any that are instantaneous, and the delay is sufficient to decant most of the hydraulic fluid over a wide area. It is not possible to carry out a simple non-destructive test to guard against any such hose failure or connection leak. The usual procedure to check a hose, pipe, or connection, is to test to a pressure above its normal working condition, which can precipitate a failure rather than anticipate it.

Due to a significant change in the physical properties of most metals and plastics at cryogenic temperatures, special care has to be taken in the choice of materials of construction of the equipment subjected to these temperatures. However, when operating with liquid oxygen, an additional product characteristic has to be considered in the interest of safety, and that is combustibility.

Oxygen is not combustible, but, it promotes combustion at a lower temperature, and more rapidly than normal when the surrounding atmosphere contains more than 20.9 per cent oxygen by volume. These effects intensify proportionately with increase of oxygen concentration, pressure, and/or temperature. Particularly hazardous are porous materials that will absorb liquid oxygen, oils and greases, etc. Many commonly used materials not normally combustible in air may burn in pure oxygen or oxygen enriched atmospheres. Special precautions are therefore essential when using equipment in service with either gaseous or liquid oxygen if fires or explosions are to be avoided (1). It is not generally realized that in conjunction with fuel substances, liquid oxygen is a more powerful explosive than TNT.

It will be appreciated therefore, under conditions of oxygen enrichment, which is not unknown in the product operating compartment of a cryogenic tanker, a high pressure fluid leak from an hydraulic system can be potentially very hazardous by providing a fuel substance, should an ignition source be in the area, or by impact ignition.

From the above it will be seen that in an hydraulic system for operation in liquid oxygen service it is essential a fire resistant fluid is used as the minimum safety measure. These fluids are available commercially with varying degrees of fire resistance, and suitability. The most acceptable in terms of fuel substance or lack of it, is a water/ethylene glycol mixture, but at a considerable penalty to the system, hydraulic units must be derated at least 50 per cent and the life expectancy can be cut by 80 per cent. This imposition increases the capital cost possibly by 25 per cent, and maintenance costs considerably, by the reduced useful life of the unit. There are other fire resistant fluids available, phosphate esters are probably the best in terms of fire resistance, lubrication and anti-corrosion properties, but handling requires a degree of protective clothing and special storage facilities are recommended, and they will burn in an oxygen enriched atmosphere. A further disadvantage of fire resistant fluids is cost, which can be double that of water/ethylene glycol, and five or six times that of the conventional antiwear petroleum oils.

(2) RELIABILITY

My opening remarks under the heading of Safety also set the scene for the importance of reliability, which within this concept could reasonably be termed an adjunct of safety.

It is not uncommon for a vehicle to be despatched to a customer six or seven day's travelling time from its base depot, often passing through one or more countries on the way. Nearer home for instance the trip from Glasgow to Dounreay is of three days' duration for a 24 000 litre capacity tanker. Under these circumstances the importance of reliability cannot be over-emphasised.

The method of pumping cryogenic liquids by an external high speed (10 000 r/min) low mass single stage centrifugal pump was introduced into the UK from the USA approximately 20 years' ago. Before that time the UK practice was to use a submerged multi-stage centrifugal pump operating at 3500 to 4000 r/min with hydraulic and external electrical drive. The hydraulic drive unit is shown in Fig 3. The original US electrical equipment comprised crude drip proof units of standard industrial design, by present day codes of practice unacceptable and totally unsuited for severe weather conditions. At that time a UK company introduced totally enclosed fan cooled (TEFC) alternative equipment suitable for liquid oxygen service and dimensionally interchangeable. The design at that time was quite unique, anticipating by many years proposed legislation and codes of practice later introduced for the safe transfer of cryogenic liquids, including liquid oxygen. It was the first custom built electric mobile cryogenic liquid pumping system introduced conforming to present day standards of safety and reliability. With considerable progress also being made by the cryogenic pump manufacturers in improving the characteristics of the cryogenic pump it was also possible to reduce the operating speed of the pump to 8000 r/min, and lower Fig 5 shows the major electrical units.

Time does not permit a full analysis of the design, however it may be of interest to report on one of the features of the motor. One cause of failure in the earlier USA equipment was the collapse of the front end bearing of the motor caused by cold creep from the cryogenic pump. Before the cryogenic pump can be brought into operation it must be cooled down to product temperature by circulating cold liquid through it for approximately 10/15 minutes. Cold is of course conducted back through the pump carrier frame to the motor end shield and along the shaft to the bearing and bearing housings, as shown in Fig 6. The lubricating grease in the bearing housing although suitable for an operating temperature of -40'C, froze solid, and on start up the bearings disintegrated. In the UK design, a heat sink with an interface was provided to reduce the cold flow and to provide a reservoir of heat in the bearing housing. To prove the design liquid nitrogen at -195.8'C (77.35 K) was circulated through the pump for 24 hours and in that time the bearing and lubricant temperature did not fall below 0'C at an ambient temperature of 7'C. One cryogenic pump manufacturer has introduced a thermal insulation system on the pump unit. This not only complements the electric motor but is of considerable value when fitted to an independent gearbox or hydraulic motor.

The hydraulic system as previously explained, normally comprises a pump driven by the vehicle engine through a Hardy Spicer lay-shaft at a speed of 1350 to 1500 r/min to produce a fluid operating pressure in excess of 200 MPa (200 bar). It is difficult to be more precise in a general way, with hydraulics, as you are aware, there are so many alternative types of unit that can be employed to achieve the end requirement which in this application you will remember is 8000 r/min absorbing approximately 45 kW. There are hydraulic units available capable of operating above 8000 r/min, but limited to 20 or so kW, there are also units available capable of greater than 90 kW, but limited in speed to approximately 6000 r/min. Neither of these systems produces the conditions of pressure and flow of the product required by the majority of customers in their process. Fig 7 shows the affect of cryogenic pump speed on performance. One of two solutions can be adopted to overcome this problem. The most favoured is to introduce a gearbox between the hydraulic motor and centrifugal pump to achieve the necessary increase in speed. The alternative to fulfil the customer's requirement at a lower product pump speed is to introduce a multi-stage cryogenic pump. This is undesirable, because it will be seen that penalties are incurred by additional weight, size, and costs, in terms of time, and heat introduced into the product. As previously shown, there are a limited number of submerged centrifugal pumps in service which totally eliminates cooling down time, but because of other influences they are not generally considered a viable proposition and are being phased out of service.

In every mobile hydraulic system the number of high pressure, joints, connections, and hoses, is considerable, and unlike the electrical joint or connection, are a potential source of failure. Hydraulic hoses are more susceptible to physical damage, age, temperature deterioration etc. than the equivalent electric cable. High pressure joints and connections are more sensitive to the inherent vibration, and relative movement experienced with heavy commercial road vehicles, than electrical connections. A simple Megger, or other test on an electrical circuit, will detect a potential failure without causing additional damage, whereas on the hydraulic system a pressure test aggravates and probably precipitates a failure. In my view to achieve a comparable degree of reliability and vehicle availability with the electrical system, the hydraulic system requires a greater expenditure of manpower and capital.

(3) EFFICIENCY

Some 25 years ago in the early days of the introduction of mobile cryogenic liquid pumping systems, economic, financial and industrial influences were quite different from those of today. Fuel costs, transport legislation, road taxes, etc., now have a much greater significance in road tanker operation than previously. In terms of power consumed to work done, efficiency in those days was low in the list of priorities. There was an abundance of cheap power available in the form of the vehicle engine or the customer's electrical supply, and providing the operation was successfully completed, it was far superior to the pressure decant method it replaced. Not only in today's climate of the conservation of energy, but also the introduction of new manufacturing processes requiring higher pressures and flows against the limited power available, efficiency is now of much greater importance. The electrical, hydraulic, and cryogenic equipment manufacturers have all introduced new and improved designs capable of efficiencies far superior to that previously thought possible. In spite, however, of improvements in hydraulic units, hose material, and manufacture, the number of bends, valves, controllers, filters etc., contribute to a loss of power from source to application greater than that in the transmission of electrical power. Fig 8 is a visual representation of the approximate power losses in both systems, and it will be seen how the higher efficiency of

© IMechE 1985 C273/85

the hydraulic pump and motor is negated by the high transmission losses. Fig 8 does not take into account the greater fuel consumption incurred by the higher r/min of the vehicle engine in the hydraulic system.

With electrical installations, providing the distance between power source and power application is known, power loss is a simple, accurate calculation. On the other hand, the hydraulic equipment poses more of a problem for the designer endeavouring to accurately determine the power available at the cryogenic pump. The design of an hydraulic installation needs careful study to ensure that necessary operating functions are realized without introducing power wastage. A surprisingly high proportion of hydraulic installations are relatively inefficient because of lack of understanding of design requirements, the introduction of conditions leading to high energy losses, and excessive fluid friction. It is under these conditions that the improvements achieved by the equipment manufacturers can be completely lost.

Efficiency can also be considered in the context of the power/weight ratio affecting the vehicle payload. With the diversity of hydraulic systems available, size of gearbox, volumetric capacity of fluid reservoir and system, etc., any assessment of weight distribution must be inaccurate, or based on an actual installed system after the event. This of course imposes a penalty on the tanker designer trying to determine axle loading to achieve a weight distribution to satisfy relevant road transport legislation. On the other hand the weight and power of an electric system is much more predictable resulting in accurate weight distribution forecast early in the design stage of the vehicle.

(4) MAINTENANCE

When a management policy making committee is considering new equipment, or a refurbishing programme, and the subject of electric and hydraulic systems is under discussion, the most common criticism of the electrical system is that installation and maintenance requires personnel with electrical training, difficult to obtain and expensive, whereas it is claimed hydraulic equipment is much simpler, it can be installed and maintained by any spare personnel available on the shop floor at no extra cost. This is a totally fallacious concept of the skills required not only to locate and install the equipment in a safe and efficient configuration, but to achieve a successful, simple, high pressure joint. The installation and maintenance of hydraulic equipment, particularly in the hostile environment of the cryogenic tanker application, requires specialist skills and expertise not normally found in a general fitting workshop. It is not uncommon for a company deciding on hydraulic equipment to ignore this basic fact and consequently experience a long, painful and expensive learning curve to improve personnel skills to an acceptable level.

Although signs of failure or impending failure in a hydraulic system are very evident by the adjacent presence of hydraulic fluid, to precisely locate the fault it is normally required to pressurise the system. Unless very carefully controlled, the test can be counter productive, particularly if the test pressure is excessive and if that excess pressure is not recorded and checked against the design parameters of the equipment. Conversely safety checks on electrical equipment are nondestructive and can normally be carried out with the equipment in a passive condition, should a malfunction occur during pumping the failsafe control system will come into operation, identify and locate the fault. Visual inspections are very often sufficient to detect a possible source of failure, and if an electrical cable is suspect, a replacement can easily be identified for power capacity, fitted, continuity checked, and is immediately available for service. On the other hand, should a hydraulic hose fail, the system must be drained, a replacement hose obtained (which incidentally must have been stored with a great deal more care than the electric cable), end connections fitted, the assembly pressure tested, installed in the system, filled with new or carefully filtered fluid, the system purged of air and finally the complete system pressure tested before it is ready for service. It will be appreciated, a much more time consuming exercise than simply replacing an electric cable.

A further consideration in the maintenance field is the gearbox. Normally both systems incorporate a gearbox. On the electric system it is located between the vehicle engine and the alternator with an input speed of 1350 r/min, a ratio of 2.76:1 giving an output speed of 3700 r/min, not a very special or arduous duty. It is also usually integral with the alternator so that should an alternator be replaced, the gearbox is automatically replaced without any stock control requirement or influence. The gearbox on the hydraulic system is usually a separate unit located between the hydraulic motor and centrifugal pump, this too may have a similar ratio of 2.76:1. There is however a fundamental difference. The input speed of this gearbox would be in the region of 3000 r/min with an output speed of 8000 r/min, a very special piece of expensive equipment.

CONCLUSIONS

Cryogenics is a young and high technological industry and as such has generated a considerable interest in research, and coupled to an innovative attitude has been responsible for many changes in our life style. At the beginning of this paper I listed four alternative systems of power transmission, two of which were only in limited use, the third in greater use and the fourth system, electrical, the most popular, and I think my previous comment shows why.

I cannot comment on the USSR. In the Far East however the two most popular methods are the electric and the pressure decant, the latter requires a high pressure tanker approximately 35 MPa (35 bar) working pressure to transfer the liquid from the tanker into the customer's storage. Possibly it was because of the later development of a large scale cryogenic industry in this area, and being given the opportunity to assess all the alternative mobile cryogenic pumping systems without massive capital investment, enabled them to by-pass the alternatives to the self generated electrical system previously mentioned.

The continual endeavour for improvement is reflected in the liquid supply industries and both electrical and hydraulic systems are under continual review. One international company recently visited, is currently renewing its interest in hydraulic systems with allegedly sufficient success to continue its investigations, with new design units which are claimed to operate at lower speeds and higher kW output with a hydraulic fluid they assert is acceptable for use in an oxygen environment. It will be of interest to learn the success achieved by the new equipment on prolonged operational service.

The hydraulic industry, within the context of the subject of this paper, is also investigating the application of failsafe systems using LED's to identify and predict a malfunction, similar to the method used in the electric system. The integration of electronics and hydraulics has been common practice in the highly sophisticated aeronautical industry for many years. In the less sophisticated road transport application of hydraulics however, the introduction of separate specialist skills, with maintenance facilities demanding a clean dry atmosphere, would be expensive in terms of time, personnel, space, and of course, money.

Bearing in mind the relative short period electrics have been in service in the mobile cryogenic pumping field, considerable improvements have been introduced into the equipment. For instance the current electrical system has doubled its power output, increased the number of safety monitoring devices, all within the existing framesize, with no increase in total weight, or cost in real terms.

Also with the advantage that can be derived from the extremely low temperature, interest is being revived in the submerged pump and electric drive motor. Under these conditions the electric drive motor can be reduced in size and weight by as much as 50 per cent, and also enables the pump manufacturer to introduce improvements in the product pump design.

With this application of electrics still in its infancy, it is probable further advances will be made to consolidate its position in this field. On the other hand, the hydraulic transmission of power was one of the earliest developments in mechanical engineering, and at the end of the 19th century it was in very common use. It is therefore difficult to envisage any new innovation of hydraulics that has not been fully developed over the past 100 years materialising, and changing the present situation, unless of course a hybrid of electrics-electronics-hydraulics-pneumatics, is introduced, with all the attendant new problems that would have to be solved.

Finally I thank Walter Jones and Company (Engineers) Limited of Sydenham London for the support and assistance they have so generously given to me in the preparation of this paper. My thanks are also due to BOC Gases Division in allowing me to make use of their slides and other information.

REFERENCES

(1) Cryogenics Safety Manual. A guide to good practice. The British Cryogenics Council.

© IMechE 1985 C273/85

Fig 1 Typical 24 000 litre cryogenic road tanker equipped with electric pumping system

Fig 2 Single stage cryogenic centrifugal pump and electric motor

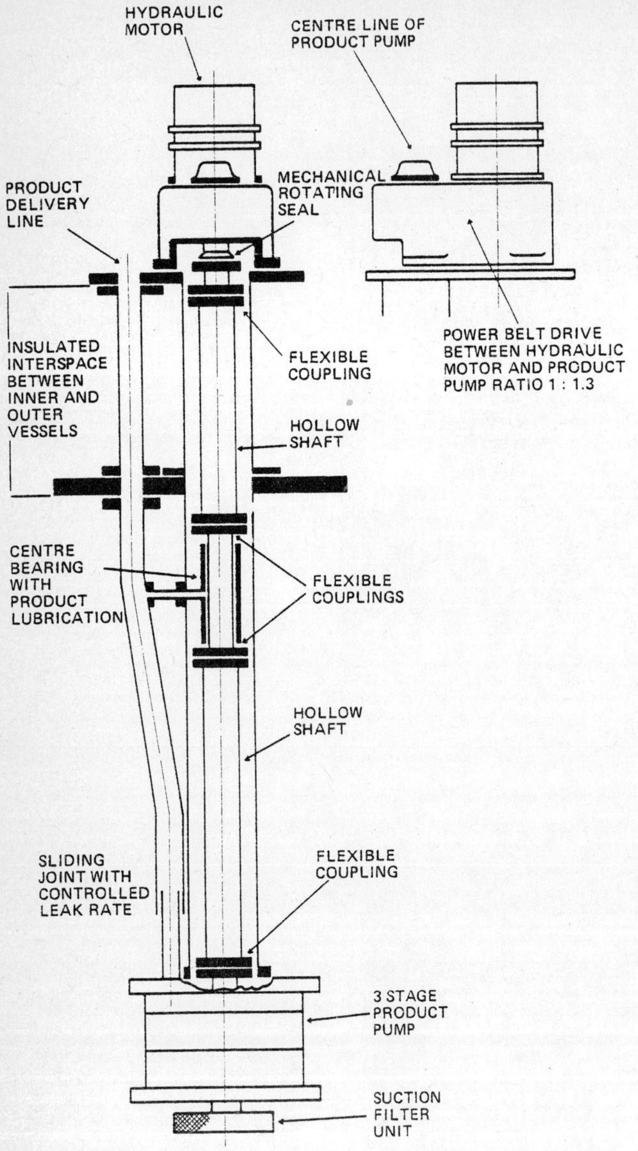

Fig 3 Multi - stage cryogenic centrifugal pump with hydraulic drive

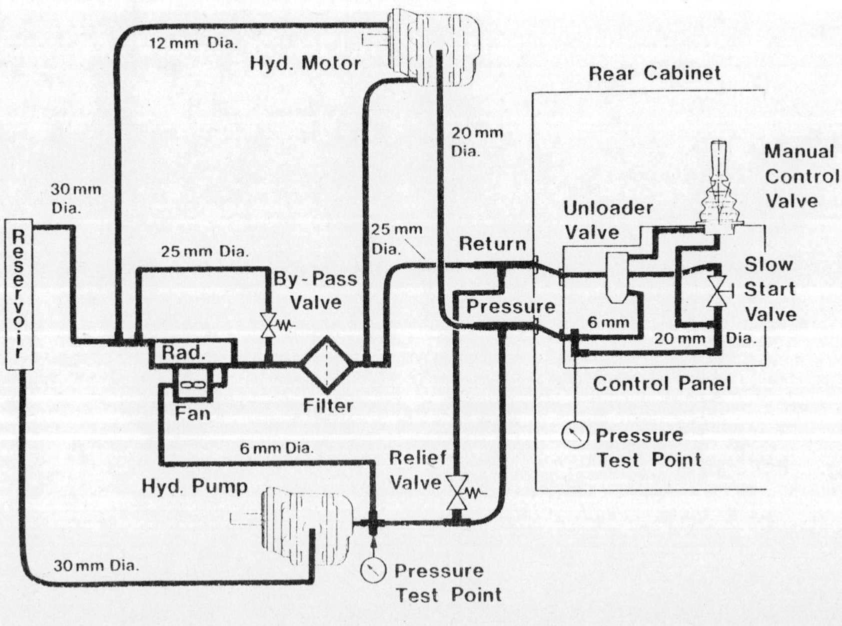

Fig 4 Typical hydraulic layout for mobile cryogenic pumping system

© IMechE 1985 C273/85

Fig 5 Electric alternator and drive motor

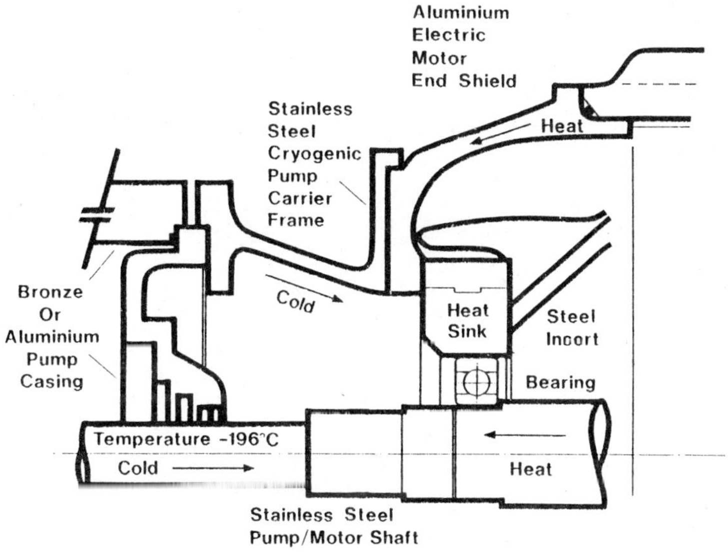

Fig 6 Illustration of cold flow from cryogenic pump to drive motor

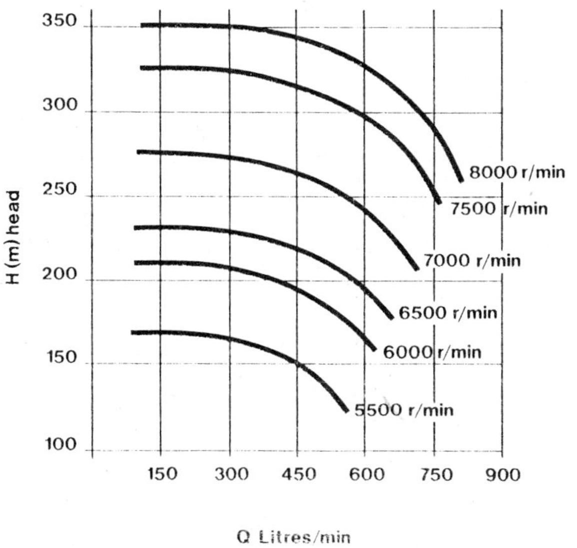

Fig 7 Effect of cryogenic pump speed on product flow and pressure

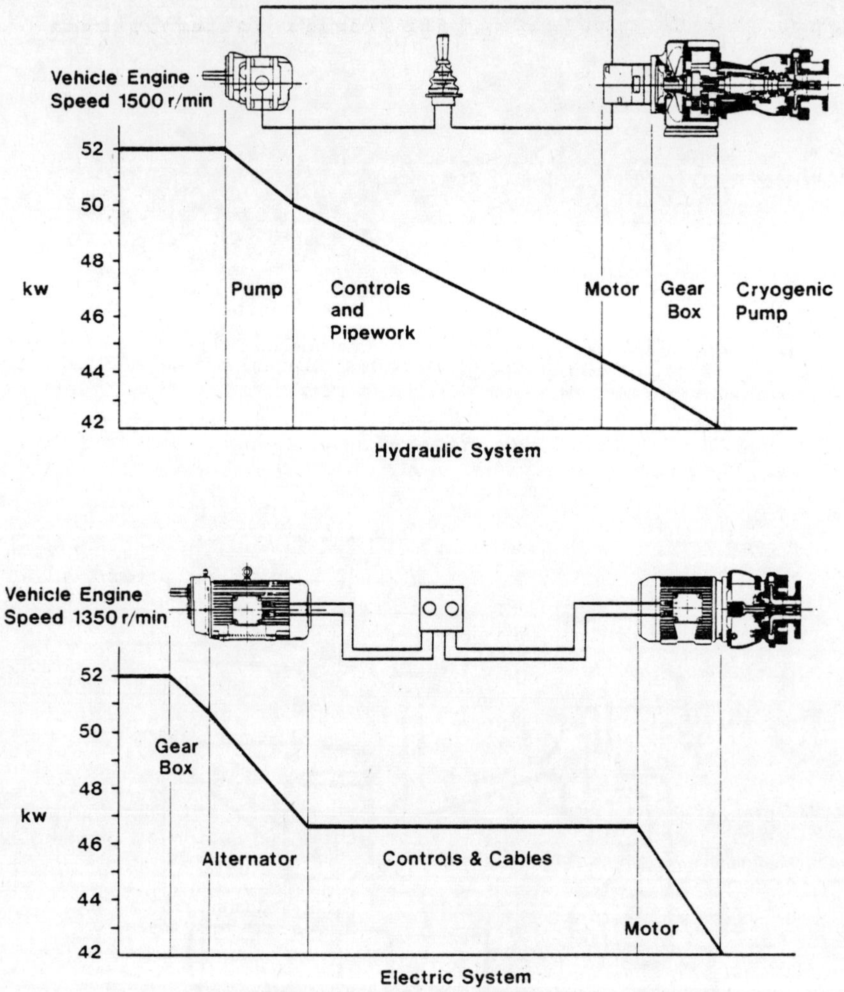

Fig 8 Visual representation of power loss in electric and hydraulic systems

© IMechE 1985 C273/85

C269/85

Automation of a wood-turning machine, hydraulic or electric?

R M GOODALL and **A H WHITFIELD**
University of Technology, Loughborough, Leicestershire

SYNOPSIS A feasibility study is described which examines the use of servo-hydraulic and electro-mechanical actuator systems for automating a wood-turning machine. Having described the system the major performance criteria are identified such that the control aspects of the two actuator system options may be assessed. A comparison based upon performance, cost and environmental suitability concludes that the electro-mechanical system is to be preferred.

1 INTRODUCTION

A manufacturer of wooden products in the UK uses a wood-turning machine to produce a number of different shapes. A feasibility study has recently been undertaken to determine possible means of enhancing the performance of this machine when producing the more irregular shapes, the major objective being to increase both throughput and consistency of the products. This study has proved to be an interesting case study of different types of actuator system: the existing machine uses pneumatic actuators, but the enhancements require higher performance actuators, and both servo-hydraulic and electro-mechanical systems have been considered. The paper presents the overall considerations relating to the specification and choice of actuator systems.

The existing machine simultaneously produces eight identical finished products from wooden blanks, and its essential details are shown in simplified form on Fig 1. The shape to be machined into the blanks is defined by a cast-steel pattern; this and the blanks are rotated on spindles about a horizontal axis at speeds of up to about 75 rpm. A tracer wheel runs on top of the steel pattern as it rotates, and this determines the vertical position of a large beam carrying eight sets of cutters. The beam is driven slowly along parallel to the axis of rotation of the blanks, such that the cutters copy the 3-dimensional shape of the pattern. Another tracer bears on to the bottom of the pattern, and this determines the position of a second beam carrying eight sanding belts (these are not shown on Fig 1). The machining operation involves two passes over the blanks. The outward pass, with the cutters alone, gives a rough and fairly rapid cut to reduce the blanks to within perhaps 5mm of the final size. The reverse pass gives a much slower and finer cut, and at the same time the sanders are brought in from below. The contact force between the tracers and the pattern is determined by pneumatic actuators and adjustable pressure valves. Another pneumatic actuator is connected to give damping of the movement of the beams.

The limitations of the machine are twofold and are especially manifest with one of the products which has a particularly irregular shape. Firstly, the somewhat crude nature of the suspension system of the beams restricts the rotational speed of the spindles - too fast a speed and the tracers jump over the corners of the highly irregular shape of the pattern as it rotates. Secondly, the rubbing action of the tracers causes significant wear of the pattern, in particular rounding off the edges. The limited life of the pattern which this causes is a particular problem because shrinkage and distortion during cooling of a cast steel pattern makes production of consistent patterns difficult.

A number of strategies for improving the machine's performance have been considered, and the chosen solution involves two distinct stages in the production process - a measurement stage and a machining stage. During the measurement stage only the pattern will be on the machine. The tracers will be allowed to bear gently on the pattern whilst it is slowly rotated, and the movements of the beam will be measured as it feeds along the length of the pattern in order to create an electronic image of the pattern, stored in a micro-electronic controller. During the machining stage, the pattern will be removed and the actuator positions will be controlled in accordance with the stored pattern data using two high performance servo systems, one for the cutter beam and the other for the sander beam. The overall scheme of additional equipment is shown by Fig 2.

The numerical procedures employed in the micro-electronic controller in acquiring, storing and retrieving the pattern data are complex, but are mentioned only superficially since the paper concentrates on the requirements of the servo system. The movements of the actuators to achieve the required performance will be described and analysed, and a specification of the servo systems derived. Outline design details of servo-hydraulic and electro-mechanical actuator systems to meet this specification will be used as a basis for a control analysis of the two types of system.

An overall comparison will be presented which takes account not only of the results of this analysis, but also of cost and environmental considerations, and general conclusions relating to this particular application will be drawn.

2 PERFORMANCE SPECIFICATION

2.1 Description of Actuator Movements

The geometrical properties of the interaction between the tracers and the pattern are complex. The shape of the cutter and sander beam tracers are different because they conform to the machining element they control - circular for the cutters and flat for the sanders. However, the details of this are not important since it is proposed to monitor and store the actual actuator movements as the pattern rotates during the measurement stage. During machining the micro-electronic controller will "replay" the stored data, although some interpolation is necessary because the spindle and feed speeds are not synchronised.

The cross section of the pattern varies along its length; the general shape is more or less constant, although the size changes. The actual pattern shape is not important: it is the actuator movements to create the shape which matter, and a series of tests has been carried out in order to quantify these movements for a typical pattern. The measured results for the cutter and sander actuators corresponding to one of the most critical cross sections is given by Figs 3 & 4. These profiles are a result of the complicated geometrical relationship between the shapes of the pattern itself and the tracers. The two position profiles are in fact quite similar, although this fact is obscured by being half a revolution offset and inverted(100 measurements per revolution were taken.). Since it is not only position that is important in specifying the actuator requirements, velocity and acceleration graphs were obtained by numerical differentiation of the results based upon a speed of 75 rpm, the maximum spindle speed. The differentiation introduces a certain amount of noise, but the general trend of the results is quite clear. The discontinuities shown clearly by the acceleration graph occur when the tracers are in the middle of some of the flat surfaces, an effect which is less pronounced for the cutter beam due to the circular nature of its tracer.

Many of the major parameters for the servo systems can be readily derived, and Table 1 lists these. Although the range of movements during machining is only in the region of 30mm, the strokes of the actuators have been made consistent with that of the existing pneumatic actuators such that the cutters and sanders can be moved well clear of the work pieces when not machining. Few mechanical details of the existing machine are available and simple measurements were made in order to determine the static forces. From these tests estimates of the beam inertias were also deduced, and these are presented as an effective mass at the attachment point of the actuators. It is clear that the combination of higher acceleration levels and greater effective mass makes the servo for the sander beam the more difficult

Table 1 Actuator System Specification

Parameter	Actuator 1 (Cutter)	Actuator 2 (Sander)
Stroke (mm)	150	150
Maximum speed (mm/s)	300	250
Average speed (mm/s)	100	100
Maximum force (N)	1600	2500
Average (steady) force (N)	1000	1700
Maximum acceleration (m/s^2)	+6, -6	+4, -12
R.M.S. acceleration (m/s^2)	4.5	5
Effective load mass (kg)	100	210

requirement. The measurements which were made, and their derived results, give the overall requirements for the servo system. However it is also necessary to determine the dynamic requirements, and further processing of the results is needed to quantify these requirements.

2.2 Further Analysis

The commands which the servos must follow are fundamentally periodic (variations in the rotational speed are small). Consequently a frequency domain approach is the obvious choice for quantifying their dynamic performance. The first stage is to perform a Fourier analysis of the position and velocity profiles, and the results of these are given in Fig 5. The fundamental component is at a frequency of 1.25 Hz, corresponding to a rotational speed of 75 rpm.

The aim of the feasibility study has been to establish guidelines for the control system design in order to meet an overall accuracy requirement of 1 mm, and for this purpose it seems sensible that the servos should respond to any harmonics with amplitudes in excess of 0.1 to 0.2 mm. Clearly a full analysis to reconstruct the servo response allowing for the gain and phase at each harmonic frequency would be a more rigorous approach, but for the purposes of establishing feasibility and for comparing actuator technologies, this simpler approach has been judged to be adequate. On this basis the cutter servo must follow reasonably up to the ninth harmonic (11.25 Hz) and the sander servo to the eleventh harmonic (13.75 Hz); position loop bandwidths in the region of 20 Hz are therefore called for.

3 CONTROL SYSTEM ANALYSIS

3.1 General Considerations

This section presents the findings of a preliminary theoretical study to investigate whether servo-hydraulic and electro-mechanical actuators are capable of achieving the dynamic requirements derived from the harmonic analysis. Since the maximum and rated forces, speeds, etc. have been accounted for in the general specification, a small signal analysis is valid for assessing bandwidth requirements. Qualitatively the normal limitations for servo-hydraulic systems are quite different from those for an electro-mechanical system, apart

© IMechE 1985 C269/85

from common features such as friction, backlash, etc. A servo-hydraulic system can apply force very directly to its actuator since the high pressure of operation means that the associated mass of hydraulic oil to be moved is small, but the compressibility of the oil and the corresponding resonance with the mass of the load must be carefully considered. By contrast, in an electro-mechanical system, the resonance effect due to the torsional stiffness of the lead screw is usually at a much higher frequency, but the inertia of the motor's armature limits the actuator's ability to transfer energy to the load at high frequencies.

The following sub-sections give details of small signal models developed for each of the systems, and their stability based upon frequency responses of the models are assessed. The relevant "imperfections" mentioned above are modelled in each case.

3.2 Servo-Hydraulic System

Responses from manufacturers of servo-hydraulic systems have indicated that 40 mm diameter actuators working off 70 bar with 40 l/min servo-valves are appropriate. Fig 6 shows a block diagram model; it employs many of the commonly accepted features of such systems (1), and includes the oil column resonance (the natural frequency of which is in the region of 75 Hz) and the pressure gain of the servo-valve, the latter being based upon 5% of maximum current causing maximum pressure against a locked load. Table 2 gives values for the model parameters. The damping coefficient has been calculated to give 0.1 damping factor in the resonant mode, and the dynamic response of the servo-valve is represented by a second order transfer function with a natural frequency of 100 Hz.

Table 2 Parameters for servo-hydraulic model

Servo-valve flow gain	K_q	0.033	$m^3 s^{-1}/A$
Actuator cross-sectional area	A	1.1×10^{-3}	m^2
Flow/force constant	K_p	4.3×10^{-9}	$m^3 s^{-1}/N$
Oil column stiffness	K	46×10^6	N/m
Oil column damping	C	20×10^3	N/ms^{-1}
Load mass	M	210	kg
Transfer function of servo-valve dynamics	$\dfrac{1}{1 + \frac{0.8}{\omega_v}s + \frac{1}{\omega_v^2}s^2}$	($\omega_v = 600\ rads^{-1}$)	

3.3 Electro-Mechanical System

Again information from manufacturers has been exploited to derive data for the model. A low inertia "printed armature" d.c. motor rated at about 600 W has been assumed. For an electric system it is sensible to offset the static force requirement using mechanical springs, and then to "match" the referred armature inertia to the load mass: this implies a lead screw pitch of around 15 mm. Maximum linear speed is with the motor running at around 1200 rpm. This is considerably within its maximum capability of 5000 rpm, but is an inevitable consequence of the matching of inertias. The model shown in Fig 7 includes the motor parameters, its viscous drag, and also the resilience of the lead screw (although the natural frequency is in the region of 200 Hz). A PWM type power amplifier has been assumed, for which a total inductance of 1.5 mH is needed to limit ripple currents.

Table 3 lists the values of the parameters, with a damping factor of 0.02 in the resonant mode.

Table 3 Parameters for electro-mechanical model.

Back e.m.f. constant	K_V	0.24	$V/rads^{-1}$
Torque constant	K_T	0.24	Nm/A
Armature resistance	R_A	0.65	Ω
Armature inductance	L_A	0.0015	H
Viscous drag coefficient	K_d	7.4×10^{-4}	$Nm/rads^{-1}$
Motor inertia	J	0.0012	$mg\ m^2$
Lead screw stiffness	K	800	Nm/rad
Lead screw damping	C	3.5×10^{-2}	$Nm/rads^{-1}$
Load mass	M	210	kg
Rotary/linear gear ratio	n	0.0024	m/rad

3.4 Assessment of bandwidths

Open loop frequency responses from the models for each type of system have been calculated. These are given on the Nichol's chart of Fig 8. As a first step the responses assume a purely proportional control strategy, and have been "normalised" on the basis of a 10% proportional band i.e. 10% of the maximum position feedback (taken as ± 50 mm) to give the maximum drive signal (20 mA for the hydraulic servo-valve and 30 V for the electric motor).

The profound effect of the oil column resonance in the servo-hydraulic system is clearly shown, and it will not be straightforward to achieve a 20 Hz bandwidth with acceptable stability margins. A "notch" type filter to attenuate the resonant effect could be added, but the success of such a technique relies heavily on the resonant frequency being relatively constant, a condition which must be satisfied as temperature and loading change.

The electro-mechanical system's open loop response is altogether more straightforward, although some form of phase lead compensation is needed. The third response shown on Fig 8 includes a 5:1 phase advance transfer function chosen to give its maximum phase lead at 20 Hz; when combined with an additional gain of +4 dB it gives a perfectly acceptable response with a closed loop bandwidth of 20 Hz. The extra high frequency gain seems to introduce no difficulty because the gain margin is still wide.

Further studies, not reported in detail here, have investigated the idea of using a velocity feedback loop, since in general terms such a loop can have a higher bandwidth and the micro-controller is equally capable of producing a velocity command from the pattern data. The response in the hydraulic system has been found to have a similarly de-stabilising effect, whereas feedback from a tacho-generator in the electric system appears to give an extremely fast-acting response.

The stability analyses have been with purely linear models, although obviously non-linearities will exist in both systems. In particular the pressure/flow characteristics of the servo-valve introduce a strong non-linearity. Nevertheless the linear models will adequately serve the purposes of a broad comparison between hydraulic and electric systems.

4. COMPARISON OF SYSTEMS

The paper now draws together the performance consideration of the previous sections, and brings in other considerations such as costs and operating environment in order to make an overall comparison.

From a performance viewpoint the electro-mechanical system offers significantly greater potential, although great care will be needed at the design stage since the control requirements are demanding. A detailed control study has not yet been carried out, and further techniques could be investigated to obtain a satisfactory performance from the servo-hydraulic system, but the inescapable presence of the oil column resonance is a severe disadvantage.

Installation of the servo-hydraulic actuators on the machine will be mechanically more simple because of their compactness, although overall the system will be more bulky due to the hydraulic power pack - the electro-mechanical system only needs a simple rectifier to give a d.c. supply for the motors.

Costs for each system have been obtained from budgetary quotations supplied by equipment manufacturers. These are detailed in Table 4, but overall they are estimated to be £4000 for the electric system and £6000 for the hydraulic system. The totals are equipment costs alone; development costs are not anticipated to be significantly different apart from the need to ensure adequate performance from the hydraulic arrangement.

The company who own the turning machine are already accustomed to electro-mechanical installations, albeit with a significantly lower level of sophistication. The power electronic amplifiers for driving the motors are clearly a "new technology" for them, as would be the high quality of the equipment for the hydraulic system. The operating environment of the machine is not particularly severe, although the presence of fine wood dust must be allowed for, especially to avoid any contamination of the hydraulic oil.

Table 4 Details of Actuator System Costs

Servo-Hydraulic System	
Hydraulic power pack	£1600
Actuators	£2200
Servo valves	£1600
Hoses, pipework, etc.	£ 600
TOTAL	£6000
Electro-Mechanical System	
Actuators	£2800
Power amplifiers	£1000
Cabling, fittings, etc.	£ 200
TOTAL	£4000

5. CONCLUSIONS

The control system application which has been described is quite unusual. The levels of force required of the servo systems are relatively small, although their speed of response must be high. The studies have shown an electro-mechanical system to be the firm favourite, partly on the grounds of cost but also because of the uncertainty in performance of the servo-hydraulic system created by the oil column resilience. It must be stressed that the control requirements are very demanding, and that work is continuing to clarify fully the specification, including further studies of the servo-hydraulic system. However this does not undermine the general conclusion that electric actuators offer greater potential for the application.

6. REFERENCE

(1) NEAL, T.P. Performance estimation for electrohydraulic control systems, Moog Technical Bulletin 126, 1974.

© IMechE 1985 C269/85

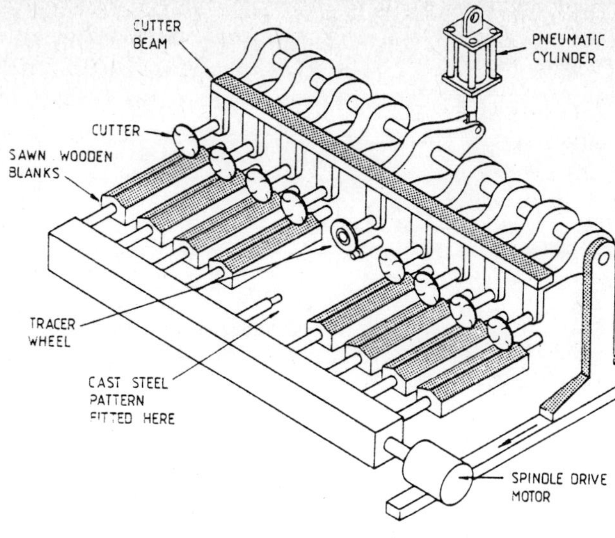

Fig 1 Scheme for wood-turning machine

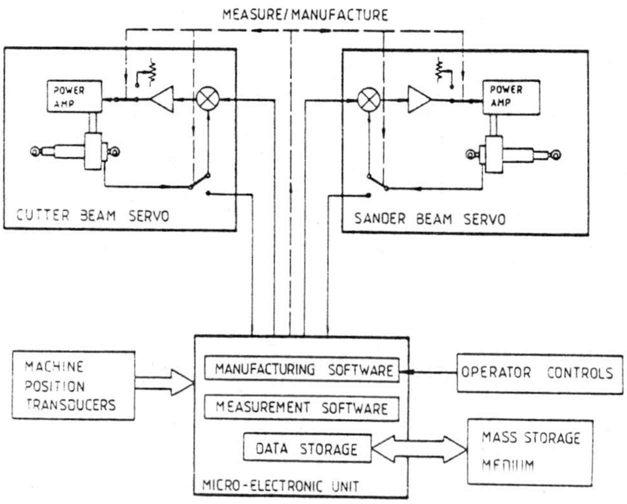

Fig 2 The overall scheme

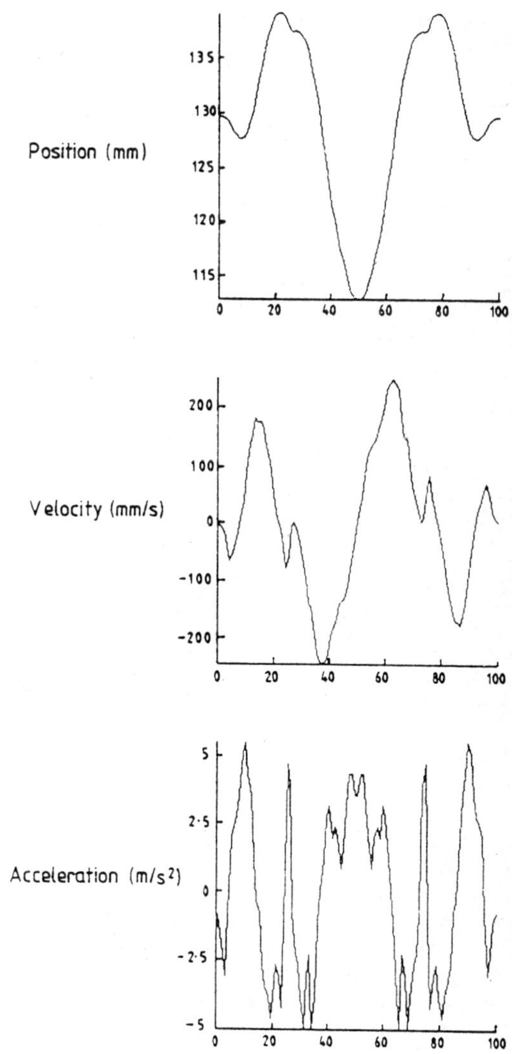

Fig 3 Cutter beam actuator

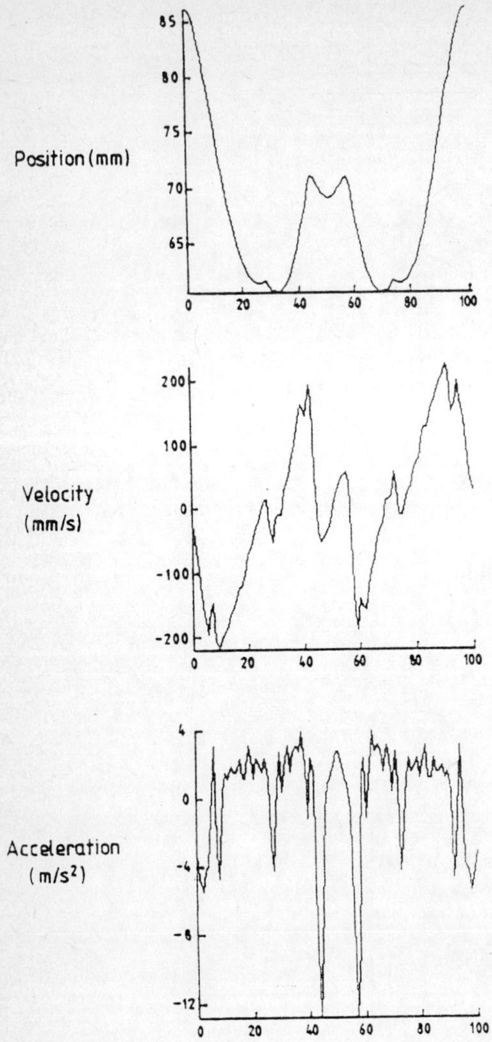

Fig 4 Sander beam actuator

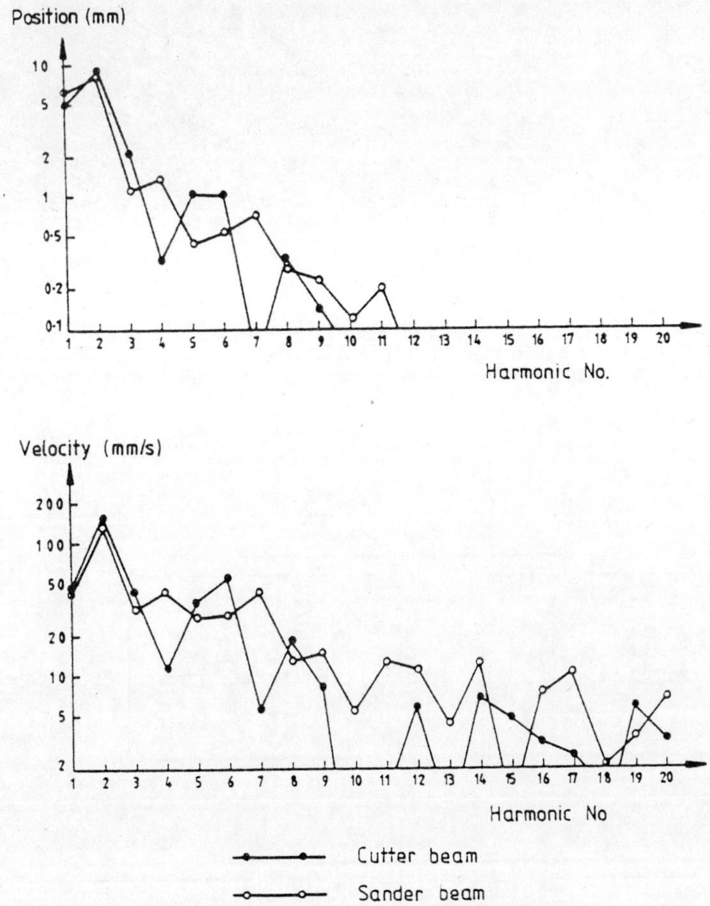

Fig 5 Results of harmonic analysis

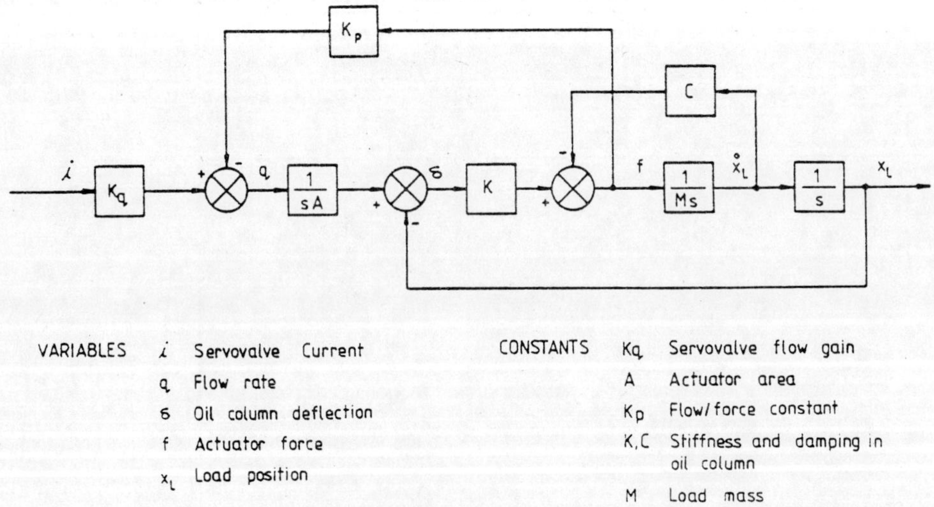

Fig 6 Servo-hydraulic system model

© IMechE 1985 C269/85

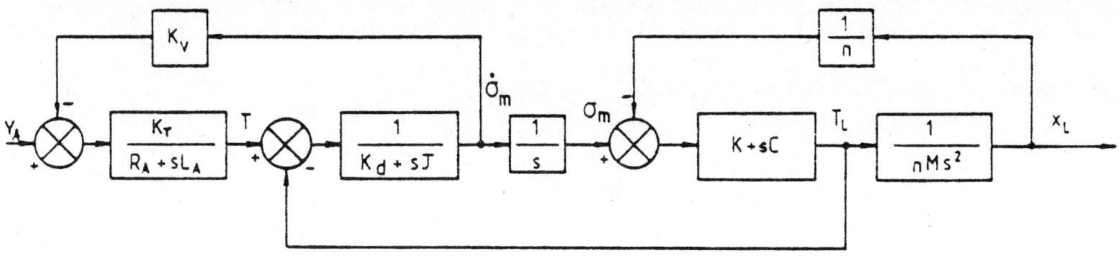

VARIABLES

v_A	Armature voltage
T	Motor torque
σ_m	Motor position
T_L	Load torque
x_L	Load position

CONSTANTS

K_V, K_T	Back e.m.f. and torque constant.
R_A, L_A	Armature resistance and inductance
K_d	Viscous torque coefficient
J	Motor inertia
K,C	Lead screw stiffness and damping
M	Load mass
n	Rotary/linear gear ratio

Fig 7 Electro-mechanical system model

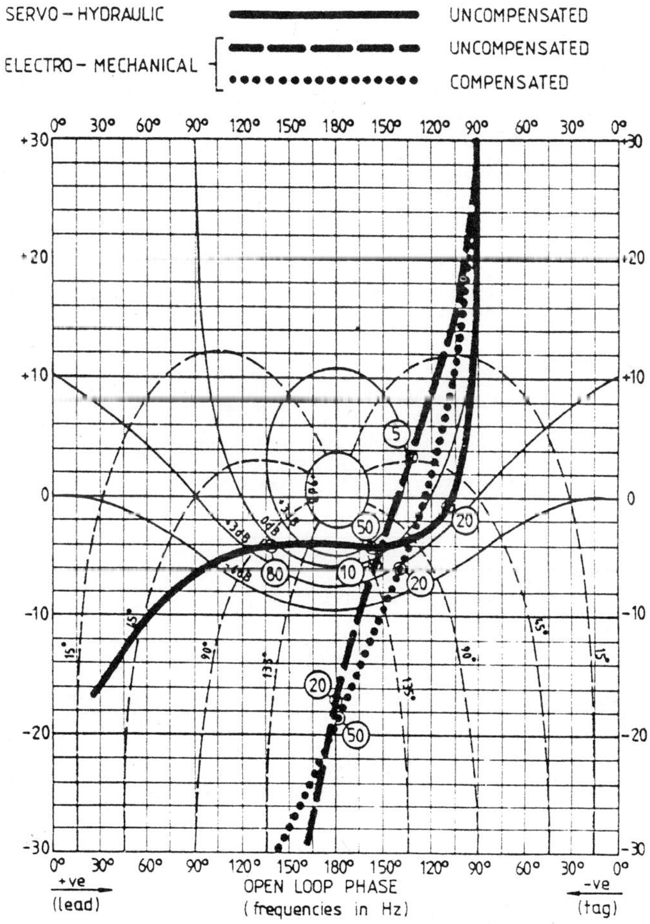

Fig 8 Open loop frequency responses

© IMechE 1985 C269/85

C276/85

The choice of a servo motor for a specific application

R FIROOZIAN, BSc, MSc, PhD and **K FOSTER**, MA, PhD, CEng, FIMechE
Department of Mechanical and Production Engineering, Aston University, Birmingham

SYNOPSIS In making a choice of the correct motor for a particular need, there are a number of factors to be considered, for example, speed of response, accuracy and dynamic error due to an external disturbance, coupled with capital costs, reliability and availability. A systematic approach is given for making the choice on the basis of a number of simplified graphs of performance, covering ceramic magnet DC motors, brushless DC motors, rare earth magnet DC motors, AC motors, stepping motors and hydraulic motors.

1 INTRODUCTION

In recent years, considerable effort has gone into improving servo motor performance, and a wide variety of servo motors are available with comparable characteristics over a range of power ratings. For the system designer the choice of a suitable motor for a particular application is quite difficult.

Hydraulic servo motors have been common in the past mainly because of their high power to weight or size ratio(1), but problems such as the effect of contamination and the necessity for a hydraulic power pack has diverted the attention of designers from hydraulic to electric motors. At the same time, electric motors capable of greater overload for reasonable periods of time have resulted in more compact electric drives, and better design and packaging have made them much more attractive to the user. The choice is from:

- Ceramic magnet DC motors
- Rare-earth magnet DC motors
- Brushless DC motors
- Stepping motors
- Variable frequency induction AC servo motors.

There are also a number of controllers available for driving the electrical DC motors from direct AC supply as:

- Single phase thyristor controlled, which produce a 50 Hz DC level pulsation;
- Three phase thyristor controlled, which produce a 150 Hz DC level pulsation;
- Pulse width modulated controller with more than 2 KHz DC level pulsation.

There has been some previous research on the performance capabilities of each of the above servomotors with different types of controller(2-6), but little work has been done to compare the performance of these motors. Some of the previous attempts for comparison are given in references (7-9). The comparison of the performance of different types of servo motors for a specific application at 1 kW were given by Firoozian, Brimson and Foster(10). The present paper extends the work of the previous one to motors of power up to 10 kW at a rated speed of 1500 rpm for a wide range of load applications. The smallest hydraulic motor available on the market was capable of producing up to 10 kW power over a rated velocity of 5000 RPM, and is used for comparison, although at lower power levels this is not comparing like with like.

The results of the analysis are given in terms of a number of parameters which have been separated out in a way that is not strictly rigorous, but which allows presentation in a series of graphs which are easy to use for design purposes.

2 NOTATION

C_m	is voltage constant of the motor
I	Armature current
I_ℓ	Load inertia
I_1	The equivalent moment of inertia of the load (kg m^2)
I_r	The moment of inertia of the rotor of the motor (kg m^2)
K_2, K_3	are the gains of the integral and proportional actions of the controller
K_4	Gain of the current amplifier
K_a	Gain of acceleration feedback
K_c	Gain of the current feedback
K_g	is the gain of the power amplifier
K_m	gain of velocity feedback
K_s	is the stiffness of the shaft connecting the load inertia
K_t	the torque constant of the motor
L	inductance of the motor
R	resistance of the motor
t_{r1}	Dynamic settling time due to the rotor inertia (sec).
t_{r2}	Dynamic settling time due to a unit

load inertia (sec/kg m^2).

t_a The total settling time (sec).

t_{s1} Saturated settling time due to rotor inertia for a unit change in velocity (sec/(Rad/sec)).

t_{s2} Saturated settling time for a unit inertia and a unit change in velocity (sec/kg m^2/(Rad/sec).

T_2, T_3 time constants of the lead-lag network

T_m The motor torque

T_s The shaft torque

V is the voltage of the power amplifier

V_e is the output of the current amplifier

V_p input to the velocity amplifier

V_v output from the velocity amplifier

ω_o Velocity variation (Rad/sec).

ω_m rotating speed of the motor

ω_o is the rotating velocity of the load inertia

3 THEORY AND PERFORMANCE CRITERIA

A servo feed-drive system consists of three sub-units;

(i) the controller
(ii) the servo motor
(iii) the load mechanism.

A mathematical model is developed for each of the sub units by employing linear or linearized relations. The details of the modelling of each of the sub-units are given in references (10-11) and are summarised in Appendix 1. In any control system employing servo motors, two important dynamic characteristics must be considered;

(a) the effect of a command signal, and
(b) the effect of external disturbances such as an external torque.

In order to investigate the complete dynamic behaviour, an impulse input response of the system would be an ideal test[12], since it excites all the modes of the system. However, because of its impracticality this test is not normally used for servo motor drive systems. A step input can be more easily applied either as a command signal to determine the speed of response, overshoot, damping and settling time, or as an external torque to determine the dynamic and steady-state accuracy in the presence of a disturbance.

In making the comparison it is assumed that all the high performance servo motors incorporate a local velocity feedback to give improved motor stiffness, that a feed-forward integrator is used to give a zero steady-state error to a ramp input and that the amplifier contains a lead-lag network for dynamic compensation. In order to obtain the best performance, the parameters of the controller such as gain, time constant of the compensation network and the gains of feedback must be optimised to give a fast speed of response and a low dynamic error caused by an external torque. It was shown[10,11] that a damping ratio of 0.6-0.7 in velocity control gives the best performance in position control and also that this damping must be achieved at maximum value of gain in order to obtain the maximum speed of response and accuracy.

In order to facilitate the dynamic analysis and optimisation of the system a computer program package was used to predict the stability and transient response[11]. The optimisation procedure is as follows;

1. The gain of the system is first increased. This results in an increase in the speed of response and a reduction in damping ratio of the system.
2. In order to compensate the reduction in damping, a lead-lag network is introduced. The usual effect of the lag network is to introduce damping in the fundamental mode of the system but has the opposite effect on the higher mode. The effect of the lead network is to increase the damping in the higher mode of vibration.
3. The improvement achieved by a lead-lag network alone is limited, since at high gain the contribution of higher modes of vibration of the systems becomes significant. Acceleration feed-back, together with a lead-lag network, improves the dynamic performance of the system by not only introducing damping in the higher modes of vibration but also preventing their contribution being dominant in the response. Motor current is used as a quasi-acceleration feedback by most manufactuers, and by increasing the gain of the current feedback a more effective compensation and damping may be achieved. If this is adopted, practical limitations such as current limit and the speed of the current rise must be considered in the optimisation[11].

Having optimised the performance, a typical response of the system to a step input of velocity and external torque are shown in figure 1. The time of response, defined as the total time to when the output reaches 95% of its final value as shown in figure 1a may be divided into saturated response and dynamic settling times, as described in the next sections.

The effect of an external torque may be represented as the maximum velocity drop from the steady state as shown in figure 1b. The velocity variation due to the external torque eventually becomes zero due to the feed forward integrator in the system. The results of the analysis were substantiated by some experimental tests which are given in reference (11) and will not be included in the paper.

4 COMPARISON OF RESULTS AND DESIGN PROCEDURE

The choice of a motor for a specific application depends on the priority placed on certain criteria, including the speed of response, stiffness to external torque, size, capital cost, reliability and availability. The main object of this section is to establish a comparison of speed of response

© IMechE 1985 C276/85

and effect of external torque, so that a designer may be the more able to choose a specific motor to meet the dynamic performance requirements. The time response of a servo motor may be split into two parts:-

1) The dynamic settling time in the linear mode, where the motor is operating within its limit condition. This implies that the operating conditions and the change in the state of the system is small, so the motor operates in a linear fashion as described by the mathematical model summarised in the Appendix. Although it will not be rigorous, the dynamic settling time due to the inertia of the rotor (t_{r1}) will be separated from the dynamic settling time due to the load inertia (t_{r2}) for case of comparison.
2) The saturated settling time, where the motor is operating at its limit conditions. This implies that the motor is drawing its maximum current and therefore providing a constant torque for the initial part of a large change from the steady state. The static settling time is a function of the required velocity variation in addition to the motor and load inertia.

Figure 2 shows the variation of the sizes of motors of various types with respect to the power rating. The inertia of the rotor also follows a similar trend to the sizes of the motor as shown in figure 3. Both sets of curves are based on data from a number of manufacturers. It can be seen that as far as the size of the motor is concerned the hydraulic motor is the smallest.

In order to differentiate the effect of load inertia on the performance, the dynamic and saturated settling times are first obtained without any additional inertia load. The increase in the overall settling time due to the load inertia is then evaluated and reduced to that for a unit load inertia, and average values are given in the figure.

The overall settling time of a servo motor with a rotor inertia of I_r and load inertia of I_l and velocity change of ωc can be approximated as;

$$t_a = t_{r1} + I_1 t_{r2} + \omega_c t_{s1} + \omega_c I_1 t_{s2} \quad (1)$$

where;

t_a is the total settling time (sec)

t_{r1} dynamic settling time due to the inertia of the rotor (sec)

t_{r2} dynamic settling time due to a unit load inertia (sec/kg m^2)

t_{s1} saturated settling time due to rotor interia and a unit change in velocity (sec/(Rad/sec))

t_{s2} saturated settling time due to a unit load inertia and a unit change in velocity (sec/(kg m^2)/(Rad/sec))

I_l load inertia (kg m^2)

ω_c velocity variation (rad/sec).

A simulation using the mathematical model of Appendix 1 was carried out to find the dynamic settling time of different types of servo motors. The dynamic settling time of servo motors without any load inertia is greatly influenced by the maximum torque availability of the motor and its rotor inertia. The saturated settling time of servo motors without any load inertia due to a unit change in angular velocity is approximately

$$t_{s1} = \frac{I_r}{T_m} \quad (2)$$

where T_m is the maximum torque of the motor and I_r is the inertia of the rotor.

Figure 4 shows the saturated settling time of servo motors and Figure 5 shows the corresponding dynamic settling time. It can be seen that the hydraulic motor gives the lowest dynamic and saturated settling time compared with electrical motors. However, the rare earth magnet and brushless DC motors can become competitive with the hydraulic motor. The AC motors have a larger saturated settling time than the ceramic magnet DC motors because of lower maximum torque capability. The dynamic settling time of AC motors, however, is close to the ceramic magnet DC motors, because although DC motors are capable of higher torque, it is necessary to limit the speed of response to avoid large current surge through the armature. The stepping motors have the largest dynamic settling time, but their saturated settling time is similar to that of AC and DC motors.

The approximate saturated settling time due to an additional load inertia is similar to that for the motor inertia, i.e.

$$t_{s2} = \frac{I_l}{T_m} \quad (3)$$

Figure 6 shows this time for a unit load inertia and for a unit change in velocity.

The dynamic settling time due to the load inertia is greatly influenced by the maximum torque capability of the motor. A series of simulations were carried out to obtain the dynamic settling time for all types of servo motors at different load inertias, and the results were averaged and scaled for a unit load inertia. Figure 7 shows the results, and it can be seen that the hydraulic motor has the largest dynamic settling time. This is mainly due to the fact that because the maximum torque of the hydraulic motor is much smaller than electrical motors it is more sensitive to load inertia. The dynamic settling time of stepping motors is larger than other electrical motors due to its sensitivity to load inertia. DC motors have the smallest dynamic and saturated settling time due to their large torque capability. The dynamic and saturated settling time of electrical motors due to load inertia reduces as the power rating of the motor increases, because higher power motors are capable of higher torque. As stated previously the single hydraulic motor covered the range up to 10 kW power range and therefore the results are straight lines.

From figures 4 to 7 a designer, by knowing the load inertia and maximum velocity variation, is then able to predict the total settling time for a particular motor using equation (1).

In many applications, the variation in angular velocity due to an external torque is important. Figure 8 shows the dynamic velocity drop for a unit step input of torque and without any load inertia. It can be seen that the electro-hydraulic motor has the highest velocity drop, followed by AC and DC motors. The speed of recovery is similar to the dynamic settling time, which can be obtained from figure 5. The velocity drop usually reduces as the load inertia increases. The reduction in velocity drop for a unit load inertia and a unit torque is shown in figure 9. It can be seen that the load inertia reduces the velocity drop of the hydraulic motor considerably and to a lesser degree for AC and DC motors. The effect of torque on stepping motors is negligible, since the error is limited to a maximum of one step, and they are not shown in these graphs. The total dynamic velocity drop, therefore, can be obtained as;

$$\Delta\omega_c = (\omega_r - I_1\omega_1)T_e \qquad (4)$$

where $\Delta\omega_c$ is the total velocity drop (rpm) and ω_r is the dynamic velocity drop for a unit external torque and without any load inertia obtained from figure 8 (rpm/N.m.).

ω_1 is the reduction in velocity drop for a unit load inertia and a unit external torque obtained from figure 9, (rpm/kg m^2/Nm); T_e is the external torque (Nm); I_1 is the load inertia (kg m^2).

The total speed of recovery which is similar to the total dynamic settling time can be obtained from figures 5 and 7.

It was earlier stated that the use of acceleration feed-back in addition to the velocity feed-back improves the dynamic performance further. Figures 10 and 11 show the dynamic settling time due to the rotor and load inertia respectively. By comparing these graphs with figures 5 and 7 it can be seen that considerable improvement may be achieved for hydraulic and AC motors. A similar trend of improvement can also be achieved on the effect of external torque.

From the above analysis the design procedure can be summarised as;

1. Calculate the equivalent load inertia applied to the motor.
2. Calculate the required duty cycle and find the maximum velocity variation and its time duration.
3. Find an approximate power rating from equation P = Txω , where P is the power, T is the external torque at the angular velocity ω.
4. From equation 1 and figures 4 to 7 find the total settling time for maximum velocity variation. Compare this settling time with the required time duration of step 2. From the comparison it should be possible to see which types of motor provide the required performance.
5. From figures 8 and 9 calculate the velocity drop due to the applied torque and compare with the permissible value.
6. If from the above analysis none of the motors provide a satisfactory performance, increase the power rating of the motor and repeat from step 4.
7. If the higher power rating motor does not produce a satisfactory result then one must use acceleration feedback. Use figures 10 and 11 instead of figures 5 and 7 and repeat the search from step 4.

From the above analysis one might find that a number of types of servo motors provide the required performance. In this case the final decision might be based on other criteria such as capital cost, reliability and availability.

To compare capital costs, the complete drive unit must be considered, including the capability of controller, the number of axes to be driven, the compatibility of possible interface with micro-computer; the addition of velocity and position transducers,and so on. There are indications that for low power rating the hydraulic servo motor drives are the most expensive. Pulse-width modulated systems are the second most expensive and they are followed by thyristor controlled DC motors; AC and stepping motors. At higher power rating hydraulic motors become cheaper than electrical motors. The AC and stepping motors are likely to be most reliable because both are brushless. Hydraulic motors are reliable but require more maintenance care. The development of brushless DC motors show considerable promise for the future. Stepping motors are not readily available in large sizes. Hydraulic motors are not available in a wide range of sizes. AC and DC motors are available in almost all power ranges. The availability of rare earth and brushless DC motors is as yet limited.

5 AN EXAMPLE ON CHOOSING A SERVO MOTOR

Suppose that a designer requires to move a load inertia of 0.1 kg m^2 from 0 to 1000 rpm at maximum of 500 ms. At this rotating speed a torque of 10. N-m. is applied to the motor which could, for example, be an equivalent of a cutting force in a machine tool. Following the procedure set out in the previous section:-

1. The load inertia is 0.1 kg m^2
2. Maximum velocity variation ω_c — 105 rad/sec. The maximum required settling

© IMechE 1985 C276/85

time is ta = 0.500 sec.

3. The minimum power requirement is

$P = T_e \times \omega_c$ = 10x105, ie approximately 1 kW. .

Therefore, the first choice will be a motor with 1 kW power rating providing it satisfies the required speed of response.

4(a) Calculation of total settling time - AC motor

From figs 4 to 7 for a 1kW power rating motor the following settling time can be obtained,

$t_{r1} = 0.1$, $t_{r2} = 8$, $t_{s1} = 0.22 \times 10^{-3}$, $t_{s2} = 0.07$

substituting in equation 1 yields;

$t_a = 0.1 + 0.1 \times 8 + 0.22 \times 10^{-3} \times 105 + 0.07 \times 105 \times 0.1 = 1.66$ sec

Comparing t_a with the required settling time of 0.5 sec it is obvious that the 1 kW AC motor does not provide the required speed of response.

If we repeat the above calculation for 2,3 kW or higher power AC motor it can be found that the 5 kW AC motors with total settling time of 0.49 sec provide the required speed of response.

The dynamic velocity drop for the torque of 10 N-m for this motor using equation 4 and figs 8,9 can be calculated as;

dynamic velocity drop = (5-4x0.1) x 10 = 46 rpm.

4(b) Total settling time - D.C. motor.

From a similar analysis for DC motors it is found that the following motors will provide a satisfactory performance;

1 kW ceramic magnet DC motor controlled with 150 Hz thyristor bridge with total settling time of 0.45 sec and dynamic velocity drop of 39 RPM.

1 kW ceramic magnet DC motor with pulse width modulated controller with $t_a = 0.41$ and dynamic velocity drop of 28 RPM.

1 kW rare earth magnet DC motor with t_a = 0.27 and dynamic velocity drop of 19 RPM.

1 kW brushless DC motor with $t_a = 0.49$ and dynamic velocity drop of 16 RPM.

Because of large dynamic settling times, other types of motors are not suitable for this application.

If dynamic response is the only criterion, one can choose a 5kW AC or a 1 kW DC motor with ceramic or rare earth magnet. The final decision obviously depends on the capital cost, size and reliability.

6 CONCLUSION

In this paper the dynamic response of a number of types of servo motor most commonly used in industry have been considered. To obtain the best performance one must optimize the gain of feedback and the compensation network in the controller. The dynamic performance in each case was represented by a settling time and dynamic velocity variation. The settling time was divided into a saturated region where the motor operates in a torque-limited condition and a dynamic region where it behaves according to the linear mathematical model. A series of simulations were carried out on different types and sizes of motors and the settling times plotted in a series of graphs in such a way that a designer can predict the settling time and dynamic velocity drop to be expected for a particular motor and application, or alternatively the motor to choose for a specific dynamic requirement.

As one might expect, no one motor excels in all circumstances, and the suitability of a servo motor depends on its application. The following general comments can be made:-

1. low load inertia and low power motor,

For this type of application hydraulic motors provide the fastest speed of response. The rare earth magnet and brushless DC motors are competitive with hydraulic motors. The effect of external torque on the velocity of the hydraulic motor is somewhat higher than electrical motors. The ceramic magnet DC, AC and stepping motors have a lower speed of response than the hydraulic and rare earth magnet DC motors.

2. high load inertia, low power motor,

Under these conditions the DC motors provide the best speed of response, followed by AC, hydraulic and stepping motors. The effect of torque becomes less significant on hydraulic motors than electrical motors.

3. low load inertia, high power motor.

Here the hydraulic motor provide the fastest speed of response, followed by electrical DC and AC and stepping motor. AC motors at high power rating become competitive with conventional DC motors. The effect of torque on electrical motors is less than hydraulic motors.

4. high load inertia, high power rating

For this condition electrical motors produce a faster speed of response than hydraulic motor the AC motors become competitive with conventional DC motors. External torque on hydraulic motors has less effect.

5. using acceleration feedback.

Acceleration feedback improves the performance of all types of servo motors. The improvement is greater for hydraulic and AC motors, but the difference in dynamic performance between all types of motors reduces. In this case the saturated performance of servo motors influences the final decision more than the dynamic characteristic, so that the maximum available torque is important.

If, for a particular application, a number of servo motors satisfy for the dynamic requirements, the final decision depends on capital cost, reliability and availability.

REFERENCES

1 Cessford, S.K., "An investigation into the design of electro-hydraulic feed drive", Ph.D. Thesis, UMIST, 1969.

2. Holmes, P.G.., "The single-phase AC feed thyristor controller as a closed-loop system element for DC motor speed control", Int. J. Elect. Engng. Educ., Vol.6, pp. 279-300, 1968.

3. Abraham, L., Forster, J. and Schliephaks, G., "AC motor supply with thyristor convertors", IEEE Trans. on Ind. and Gen. Appl., IGA-2, 344, 1966.

4. Murphy, J.M.D., "Thyristor control of AC motors", Pergamon Press, 1973.

5. Lawrenson, P.J., Hughs, A. and Acarnley, P.P., "Starting/stopping rates of stepping motors: Improving and prediction", Proc. Inst. Con. on Stepping Motors and System, July 1976.

6. Barber, N.T., "The future of DC servo drives", Electric versus Hydraulic Drives, I.Mech.E. Conf. Publs., October 1983-13, London.

7. Lawrenson, P., "Switched reluctance motor drives", Electronics & Power, February 1983.

8. Ball, D., "There is no generally accepted view of this technology in ten years' time", Electronics & Power, February 1983.

9. Jones, D.H., "Matching a motor to its load using dynamic motor characteristics", Control Engineering, February 1975.

10. Firoozian, R., Brimson, J.R. and Foster, K., "The comparison of performance of servo feed-drive systems", I.Mech. E. Conf. Publs., October 1983, London.

11. Firoozian, R., "The comparison of the performance of servo motors feed-drive systems", Ph.D. Thesis, Dept. of Mech. Eng., University of Aston in Birmingham, 1983.

12. Schultz, W.C. and Rideout, V.C., "Control system performance measures: Past, present and future", IRE Transaction on Automatic Control, October 31, 1960.

KEY TO THE FIGURES

1. Ceramic magnet DC motors.
   1a. controlled by 50 HZ thyristor bridge
   1b. controlled by 150 HZ thyristor bridge
   1c. controlled by pulse width modulated with frequency of 2 KHZ.
2. AC induction motors.
3. Stepping servo motors.
4. Electrohydraulic servo motors.
5. Rare earth magnet DC motors.
6. Printed or brushless DC motors.

Appendix 1 MATHEMATICAL MODEL

A.1. Controller.

The controller of servo motors usually consists of a) velocity amplifier, b) current amplifier c) compensation network and d) power amplifier.

The dynamic equation of the velocity amplifier with velocity and acceleration feedback and a feed forward integrator can be written in Laplace transform notation as;

$$\frac{V_v}{V_p - K_m\omega_m - K_a s\omega_m} = \frac{K_2 + K_3 s}{S} \qquad \text{(A1)}$$

where, V_v is the output from the velocity amplifier

V_p is the input to the velocity amplifier

K_m and K_a are the gain of velocity and acceleration feedback respectively

K_2 is the gain of the integral action of the controller

K_3 is the proportional gain of the controller

s is the Laplace transform operator.

ω_m is the rotating speed of the motor.

The transfer function of the current amplifier with a lead-lag network may be written as;

$$\frac{V_c}{V_v - K_c I} = \frac{K_4(1+T_2 s)_4}{1 + T_3 s} \qquad \text{(A2)}$$

where; V_c is the output of current amplifier

K_c is the gain of current feedback

I is the current measured at the armature.

T_2 and T_3 are the time constants of the lead-lag network respectively.

K_4 is the gain of the current amplifier.

The equation of the power amplifier in linear form can be written as;

© IMechE 1985 C276/85

$$V = K_g V_c$$

where; V is the output of the power amplifier and K_g is the gain.

A.2. servo motors.

The voltage equation of the servo motors can be written as;

$$V = RI + LsI + C_m \omega_m \qquad (A4)$$

where; R and L are the resistance and inductance of the motor. C_m is the voltage constant of the motor.

The torque equation of the motor in linearized form can be written as;

$$T_m = K_t I \qquad (A5)$$

where; T_m is the generated torque and K_t is the torque constant of the motor.

The dynamic equation of the rotor with an inertia, I_m, and shaft torque, T_s can be formulated as;

$$T_m = I_m s\omega_m + T_s \qquad (A6)$$

A.3. Load.

The dynamic equation of a load inertia I_l and an external torque of T_l with a dynamic friction of C_l can be written as;

$$T_s - T_l = I_l s\omega_o + C_l \omega_o \qquad (A7)$$

where; ω_o is the rotating velocity of the load inertia, if the load inertia is connected to the motor through a flexible shaft with stiffness K_s then;

$$T_s = K_s \left(\frac{\omega_m}{s} - \frac{\omega_o}{s}\right) \qquad (A0)$$

Having obtained the dynamic equation of each part of the system the overall block diagram of the system is shown in Fig. 12. This mathematical model is in general form and can easily be modified for a.c, stepping and electro-hydraulic motors.

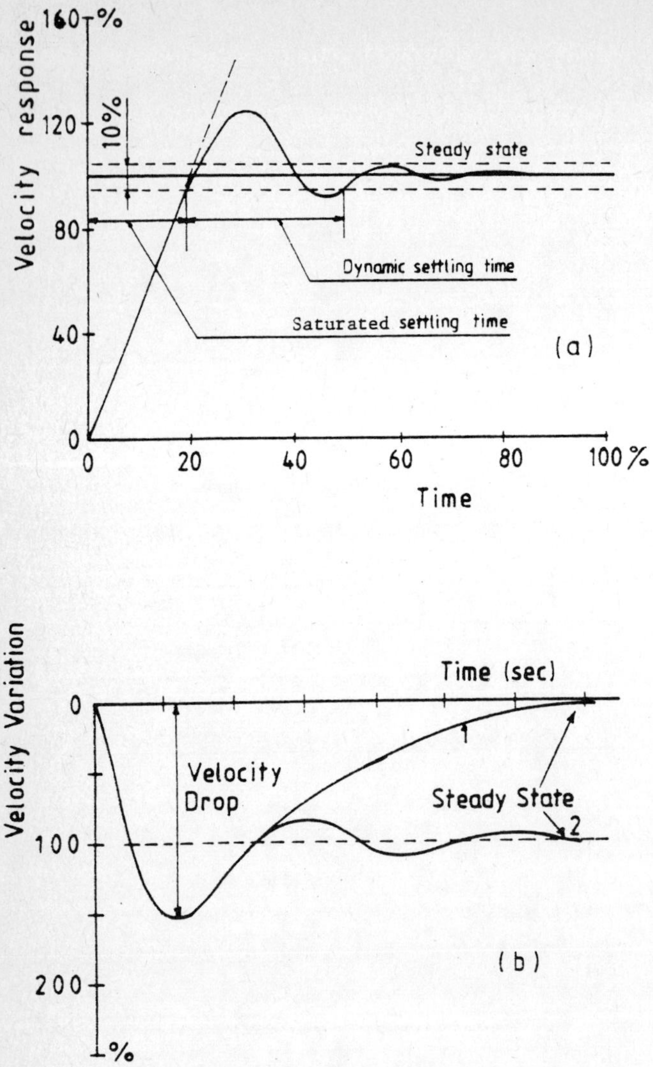

Fig 1 A typical step input response of servo motors
a Step input response
b Step external torque response

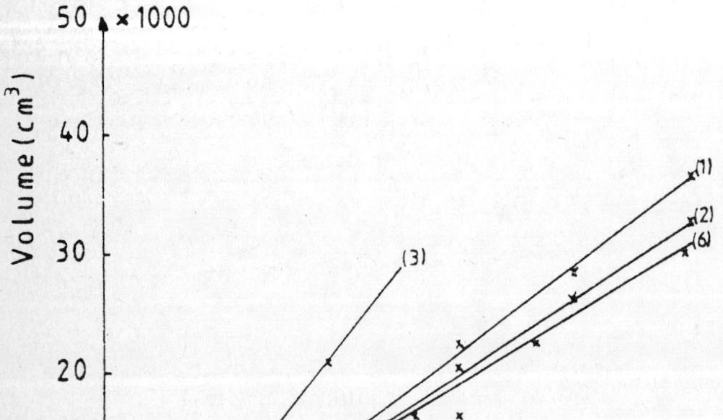

Fig 2 A comparison of volume (proportion to weight) of different servo motors

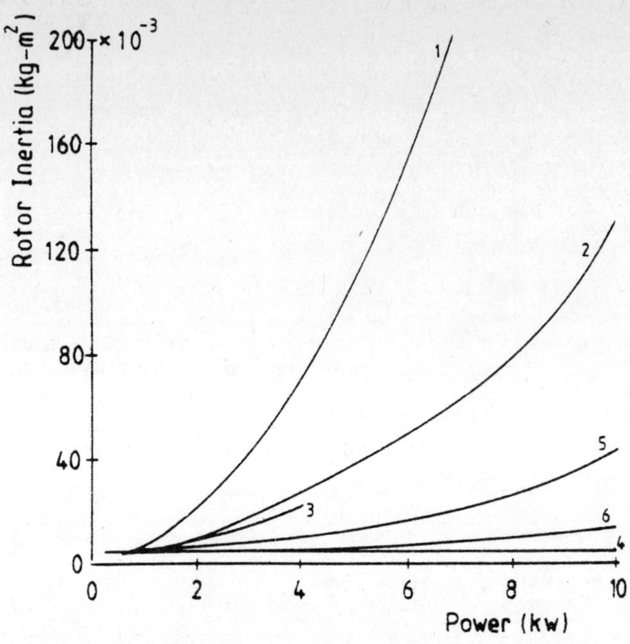

Fig 3 A comparison of the rotor inertia of servo motors

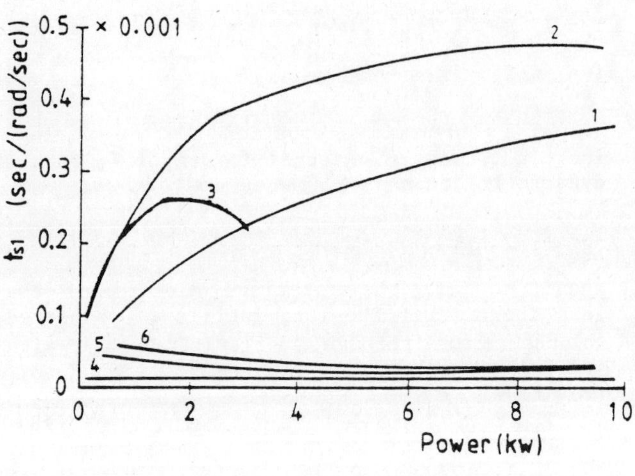

Fig 4 Saturated settling time for a unit change of velocity without any load inertia

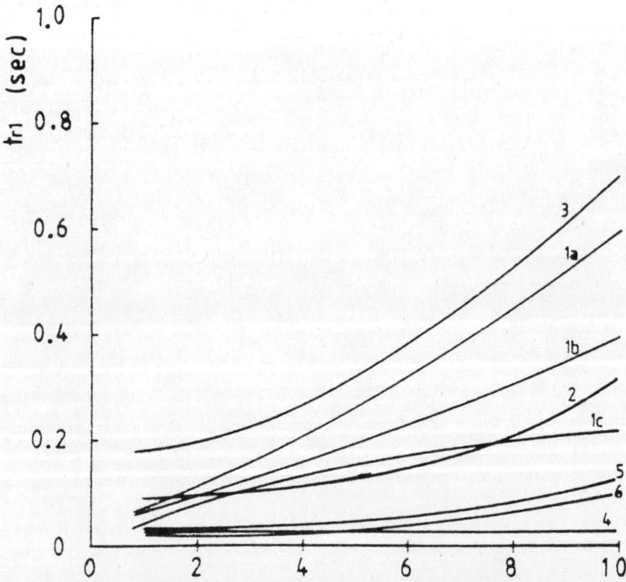

Fig 5 Dynamic settling time of servo motors without any load inertia

© IMechE 1985 C276/85

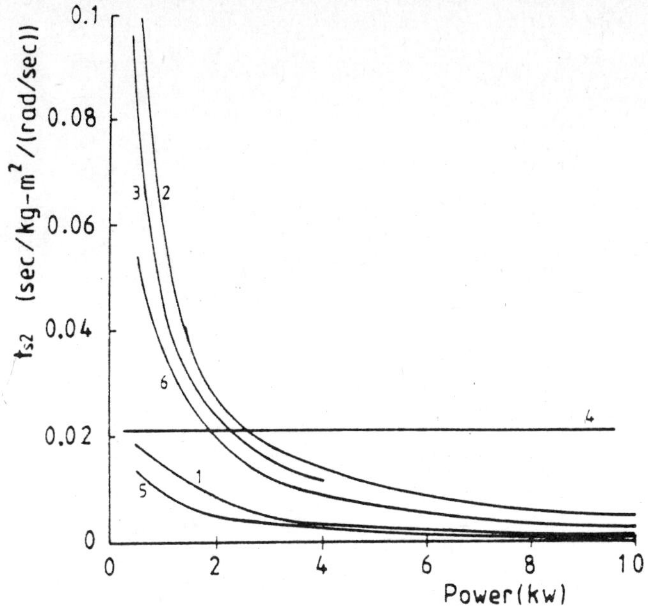

Fig 6 Saturated settling time for a unit change of velocity and for a unit load inertia

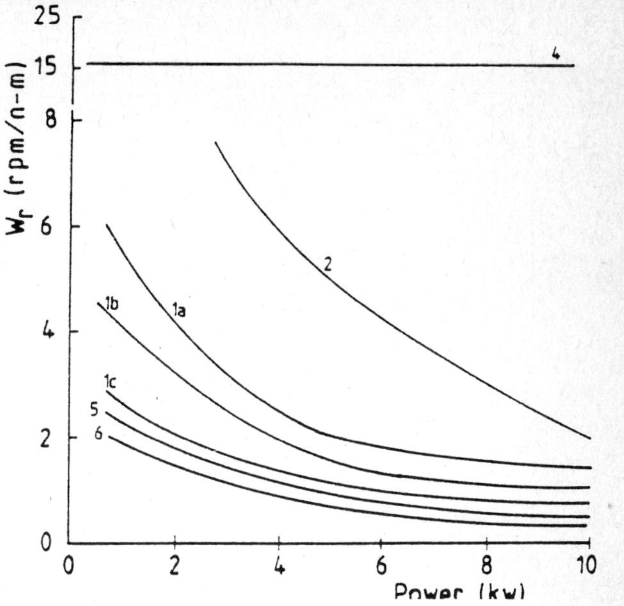

Fig 8 Dynamic velocity drop of servo motors for a unit torque without any load inertia

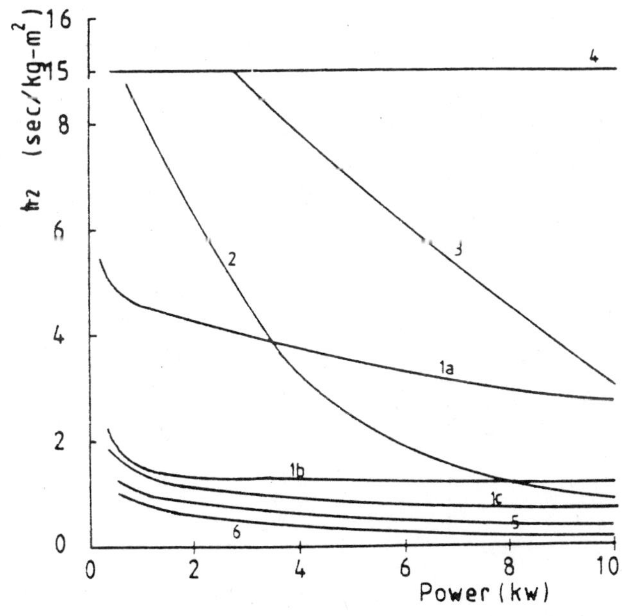

Fig 7 Dynamic settling time of servo motors due to a unit load inertia

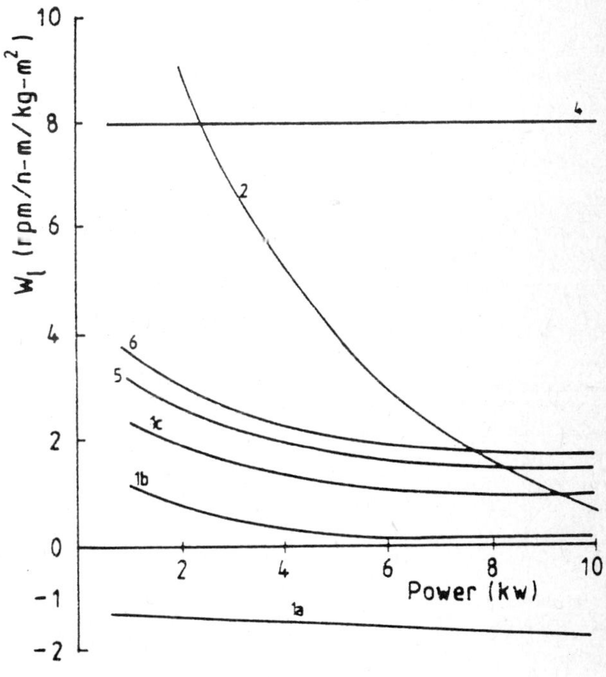

Fig 9 Reduction in dynamic velocity drop for an addition of a unit load inertia at a unit load torque

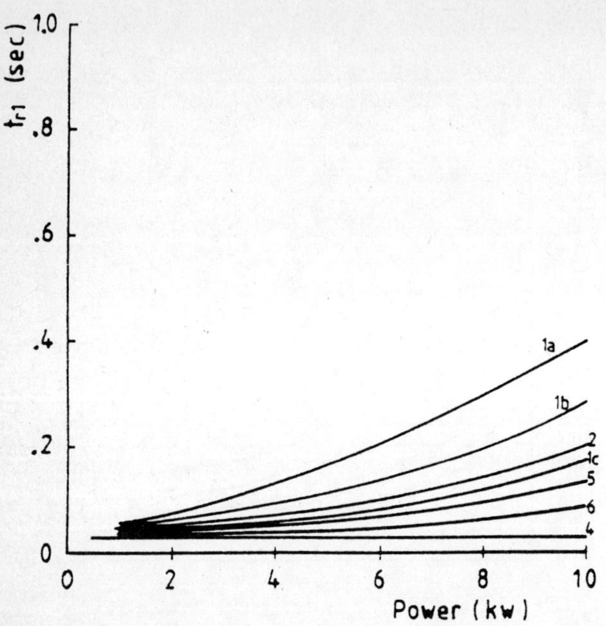

Fig 10 Dynamic settling time of servo motors without any load inertia, when acceleration feedback is used as compensation

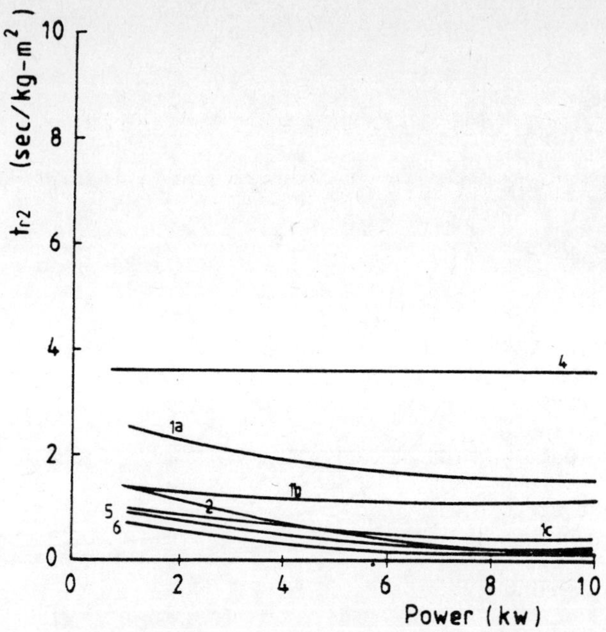

Fig 11 Dynamic settling time ot servo motors for a unit load inertia when acceleration feedback is used as compensation

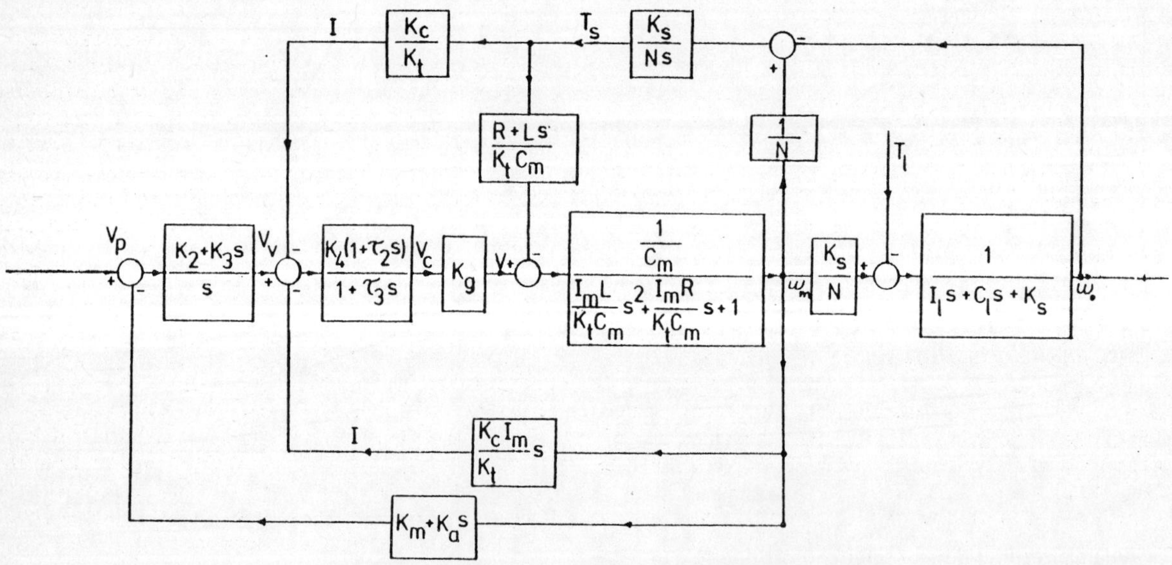

Fig 12 Linear mathematical model of the system

© IMechE 1985 C276/85

C271/85

Developments in the design and control of pneumatic linear actuators

P J BIRD, MA, CEng, MIMechE
Martonair Limited, Farnham, Surrey

SYNOPSIS The advantages and limitations of the conventional pneumatic cylinder are discussed. Recent developments in the design of pneumatic linear actuators have resulted in the production of more compact and better guided actuators. However, the most important development is the introduction of an electro-pneumatic servo control system applicable to the full range of pneumatic linear actuators. The servo control system developed by Martonair and its integration into complete control system is described together with several typical applications.

1 INTRODUCTION

For manufacturing industry where factories are equipped with a compressed air supply at 6-7 bar, there can be no doubt that the pneumatic cylinder is the most cost effective linear actuator for moderate forces and distances.

A typical double acting pneumatic cylinder used in industry with a diameter of 63mm and stroke of 320mm will provide a useable thrust of 1300 Newtons (130Kg) and will be able to complete a full stroke in less than one second. Such a cylinder will sell for a price of around £40. With air line lubrication one can expect well in excess of 5 million operations before servicing is required and even then a cheap replacement seals set will restore the cylinder to almost new condition. Any other type of linear actuator will have an initial cost of some hundreds of pounds and is likely to require much more expensive and frequent maintenance.

However, the conventional pneumatic cylinder has four main limitations:

(a) the length required for the installation of a cylinder is considerably more than twice the stroke length. For example a 63mm cylinder with a 320mm stroke will be over 800mm long when the piston rod is extended. Although this does not normally cause problems there are some applications where conventional cylinders cannot be accommodated.

(b) The cylinder is essentially a device for providing thrust and it is normally necessary to provide suitable mechanical elements to guide and locate the load being moved by the cylinder.

(c) The cylinder can only operate between two positions which must be defined by mechanical stops, provided either externally or by the piston contacting the ends of the cylinder.

(d) The motion of the piston cannot be controlled precisely, as usually the only means of control are manually adjusted air flow regulators and the pneumatic cushioning over the last 25mm or so of the stroke.

During the past few years, there have been a number of developments in design and control of cylinders to overcome these limitations and increase the scope for the application of pneumatic linear actuators. Rodless cylinders deal with the first limitation, handling modules with the second and electronic servo control with the third and fourth.

2 RODLESS CYLINDERS

Slotted tube rodless cylinders have been introduced, making it possible to install pneumatic cylinders in very much smaller spaces than in the past. The two best known types are the Origa cylinder, and the Lintra cylinder marketed by Martonair (Fig 1). Both these designs have a piston assembly connected to an external member outside the cylinder through a longitudinal slot in the barrel. The slot is sealed by means of a metal or plastic sealing strip making it possible to pressurise both ends of the cylinder in the normal way. These rodless cylinders are very much shorter than their conventional counterparts, greatly easing installation problems. For example a 63mm cylinder with a stroke of 1 metre will have an overall length of only 1,200mm.

The Lintra cylinder also goes a long way towards meeting the second limitation referred to above as the external driven member is in the form of a carriage running in Vee guides which are integral with the cylinder barrel. The Lintra cylinder therefore constitutes a complete machine element rather than simply an actuator.

3 HANDLING MODULES

In order to simplify the assembly of pneumatically actuated handling equipment, a number of companies supply ranges of modules which include pneumatic cylinders to provide the actuation as well as guidance systems to eliminate the need for separate external mechanical items.

The ranges normally include several sizes of linear modules together with adaptors to link the modules together in a variety of ways and such items as semi-rotary actuators and grippers. Figure 2 shows the main elements of the range offered by Martonair and a wide range of handling units can be assembled from these items. These modules in their standard form have adjustable mechanical stops and there are only two positions for each actuator. For many applications this is quite adequate but for some requirements such as palletising, multiple positions are required. For such applications an electronic servo control system is necessary.

4 PNEUMATIC SERVO CONTROL

To overcome the control and positioning limitations mentioned above, Martonair Ltd have been working for several years in conjunction with Loughborough University of Technology to develop a servo control system which can be applied to handling modules and also to conventional pneumatic cylinders and rodless cylinders (1). The system was introduced at the Automan '85 Exhibition held in May of this year and provides electronic control of motion with a positioning accuracy of plus or minus 0.1mm and enables actuators to achieve speeds of over 1 metre per second on longer moves.

The arrangement of a servo driven module is shown diagrammatically in Fig 3. The actuator contains a double acting pneumatic cylinder which is controlled by a solenoid operated servo valve, taking its air supply from the normal factory airline at a pressure of 4 to 7 bar. An incremental encoder is mounted on the actuator and driven by a rack and pinion from the moving element, which gives digital position information to the single axis controller.

Fig 4 illustrates the operation of the motion control process. A command is transmitted by the supervisory controller to the position controller defining the desired new position. The controller compares the present position with the desired position and causes the servo valve to move away from the mid position. Air flows into one end of the cylinder and out of the other causing the actuator to move. As the actuator moves, the controller reads the position every few milliseconds and the velocity and acceleration are computed. An appropriate signal is then sent to the servo valve to provide the necessary flow of air to and from the cylinder to produce the correct motion profile to bring the actuator to rest at the desired position.

From the application point of view one of the most important performance aspects is the total time required to carry out a movement. The mean positioning speed depends on the length of the move and the pay load. Clearly short moves will have a lower mean speed than long moves as a significant proportion of the total time is taken settling at the desired position. However, as an indication of performance, when operating with a pay load of 5 kilograms, moves having a magnitude between 80mm and 500mm are completed in times ranging from 1 second to 2 seconds.

The main elements of the servo driven modules are the programmable actuators themselves, the servo valve and the position controller.

4.1 Programmable actuators

A typical programmable handling module is shown in Fig 6. The moving element consists of a length of aluminium extrusion which is square externally and round internally. The outside surfaces have steel wearing plates which contact bearings in the fixed housing. The extrusion also forms the barrel of an air cylinder with a piston carried on a fixed piston rod, so that compressed air supplied to the ends of the cylinder cause the extrusion to move relative to the piston. The figure shows the encoder. An optional electro-magnetic lock may be fitted to increase the stiffness of the actuator once the set position has been reached. This is not a brake and is not used to control the motion of the actuator. It is only used when the actuator has been brought to a halt under servo control.

The servo control system can also be applied to conventional pneumatic cylinders and to a variety of other pneumatic actuators such as slotted tube rodless cylinders, cable cylinders and semi-rotary actuators (Fig 6). In general the only change required is the fitting of a suitable feed back device. If an encoder is used it may be mounted on the cylinder itself or on the moving machine element and may be driven by a rack and pinion, fast lead screw or other suitable mechanism. If a lock is required this can usually be more conveniently fitted on the associated machine rather than on the pneumatic actuator itself.

4.2 Servo Valves

The construction of the valve is shown in Fig 5. The valve is a 5 port design with a matched metal to metal spool and sleeve. This item has been developed by Martonair specifically for servo control and has been used for many years in pneumatic positioning equipment. The valve is operated by a proprietary proportional solenoid and provides a basically linear relationship between air flow rate and current. Although simpler and therefore cheaper than its counterparts in hydraulic servos, the valve is far more controllable than the on off valves which have been used by others in electro-pneumatic systems and this has contributed greatly to the performance noted above.

4.3 Position Controller

As is shown in Figure 3 the system includes an electronic controller for each servo controlled axis. It would be possible to use a single processor for two or more axes but separate input and output circuits are still required for each axis and the use of separate controllers adds relatively little to the hardware cost and gives advantages particularly regarding communications with the supervisory controller.

The position controller uses the Texas Instruments TI9995 16-bit micro-processor. This processor was selected because the control software was developed by Loughborough University of Technology (2) who have considerable experience of the TI9900 family of microprocessors.

The controller includes circuitry for driving the servo valve and for conditioning the position sensor output. It also has an RS232 serial port and parallel connections for a hand held non-micro-processor terminal and a PLC sequencer. The RS232 port can be upgraded to RS422 for use on a full data communications network.

© IMechE 1985 C271/85

4.4 Supervisory Controllers and Communications

The choice of supervisory controller for a system which includes one or more servo controlled axes will depend on the application. In most cases the system will also include a number of positive strop motions particularly for grippers, wrist rotation and short linear movements. A wide variety of controllers can be used but the three main options are micro-computers, PLCs with digital output and sequencers. The position controller has provision for operation with each of these.

If a micro-computer is used, connection with each position controller will be via the RS232 serial port. The set point data can be transmitted to the controller when required and this type of control is particularly suitable when frequent reprogramming of the set points is required. The RS232 link can also be used when setting up the system to adjust the control parameters to suit the type of actuator and the load.

For many handling systems, each actuator will move between a number of fixed positions which will rarely need to be reset. This type of system can be most conveniently controlled by a PLC or sequencer with digital output provided that there is provision for setting the required set points. Each position controller can be programmed with up to 31 set points and each of these can be called up as required via the parallel connection from the PLC or sequencer. The set points are programmed from the hand held terminals which have teach facilities enabling the actuator to be moved slowly to the required positions. The hand held terminal can also be used to set the control parameters.

5 APPLICATIONS

There are a wide range of possible applications for servo controlled pneumatic actuators and some of the main types can be categorised by the frequency of programme changes.

5.1 Applications Requiring Frequent Re-programming

A typical application of this type is a computer controlled saw. The requirement arose in Martonair's own factory where it is necessary to cut round bars to length for the manufacture of piston rods for pneumatic cylinders. Cylinders are produced to order with a wide range of stroke lengths and piston rods have to be cut to a corresponding range of lengths.

As an alternative to purchasing a dedicated CNC saw it was decided to build a machine using a proprietary free standing saw and a servo controlled module for the feed mechanism. The machine handles bar up to 32mm in diameter and cuts billets to length with an accuracy of 0.2mm. The module has a stroke of 600mm but the machine will cut longer lengths by using multiple feed strokes.

In order to provide a user friendly supervisory control system a small micro-computer is used in conjunction with a monitor screen. The micro-computer transmits set point information to the position controller by means of a serial link as described earlier. A menu driven programme allows the operator to input bar length, batch size, etc. quickly and easily.

5.2 Applications Requiring Infrequent Re-programming

There are many material handling applications in batch manufacturing industry where the need for changes to the programmed positions will only occur between batches, to suit dimensional variations between different components. Such applications include loading and unloading of machine tools, presses, etc. and some automated assembly operations.

In many cases an assembly of pneumatic handling modules using end stop modules for fixed movements and servo controlled modules for the movements which vary with component size, will provide a convenient and cost effective solution.

Generally the most convenient supervisory controller will be a PLC or sequencer. The provision for up to 31 pre-set positions on each servo controlled axis will usually be sufficient to cover the full range of components. The appropriate set points can then be called up as required from the supervisory controller.

5.3 Applications Rarely Requiring Re-programming

Some handling operations involve a large number of fixed positions but the sequence will rarely change. A particularly important example is palletising, where items are unloaded from, or loaded to, a rectangular array of positions. This involves the use of two axes at right angles each of which will have to move to perhaps 10 or 12 positions.

Examples are shown in figures 7 and 8. Figure 7 shows a demonstration palletiser using a servo controlled convenetional cylinder to move the pallet along the X axis. A servo controlled module provides the Y axis and the Z axis movement and gripper are end stop motions. This system is controlled by two position controllers with a Martonair M/3720 sequencer as the supervisory controller.

Figure 8 shows a practical example loading components into a vertical drilling machine. Two servo controlled rodless cylinders provide the X and Y motions and all other movements use N stop devices. Components are loaded to the machine on a 10 x 3 pallet and each component is loaded to the drilling machine and replaced on the pallet after drilling. The machine has a two-position rotary table so that the handling sequence takes place at the same time as the drilling operation minimising the overall cycle time. The system is again controlled by two position controllers and a Martonair M/3720 sequencer.

6 CONCLUSION

Development in the design and control of pneumatic linear actuators are increasing the scope for the application of these drive devices. The development of pneumatic servo control makes it possible to use pneumatic actuators for many applications which would otherwise require more expensive electric or hydraulic actuators. In particular it is believed that the system will find many applications in the material handling field where simple pick and place devices are not adequate but the full performance of multi-axis robots is not required.

REFERENCES

(1) WESTON, R. H. and MORGAN, G. A new family of robot modules and their industrial application. Proceedings of I.Mech.E. Conference 'UK Robotics Research 1984' pp. 51-57.

(2) WESTON, R. H., MOORE, P. R., THATCHER, T.W., and MORGAN, G. "Computer controlled pneumatic servo drives". I.Mech.E. Part B 1984.

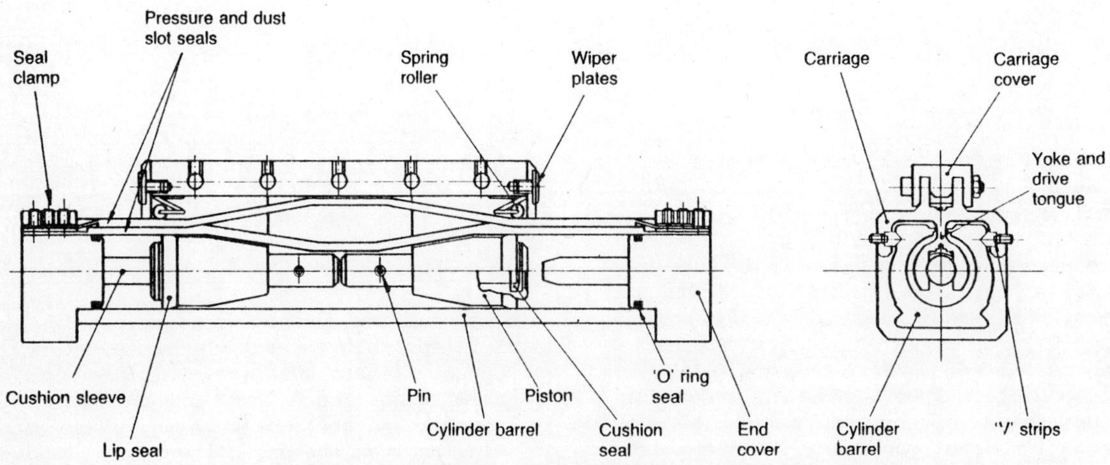

Fig 1 Sectional view of a Lintra slotted tube rodless cylinder

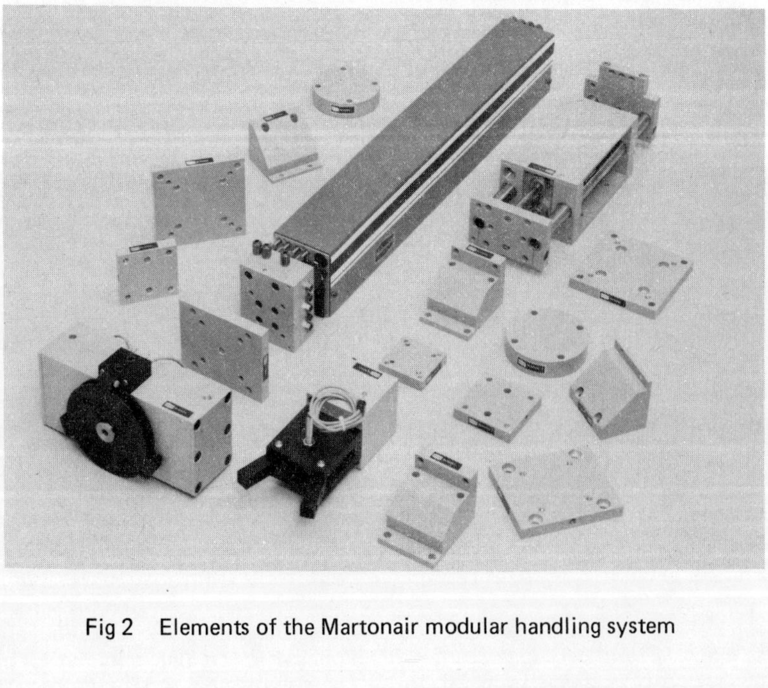

Fig 2 Elements of the Martonair modular handling system

© IMechE 1985 C271/85

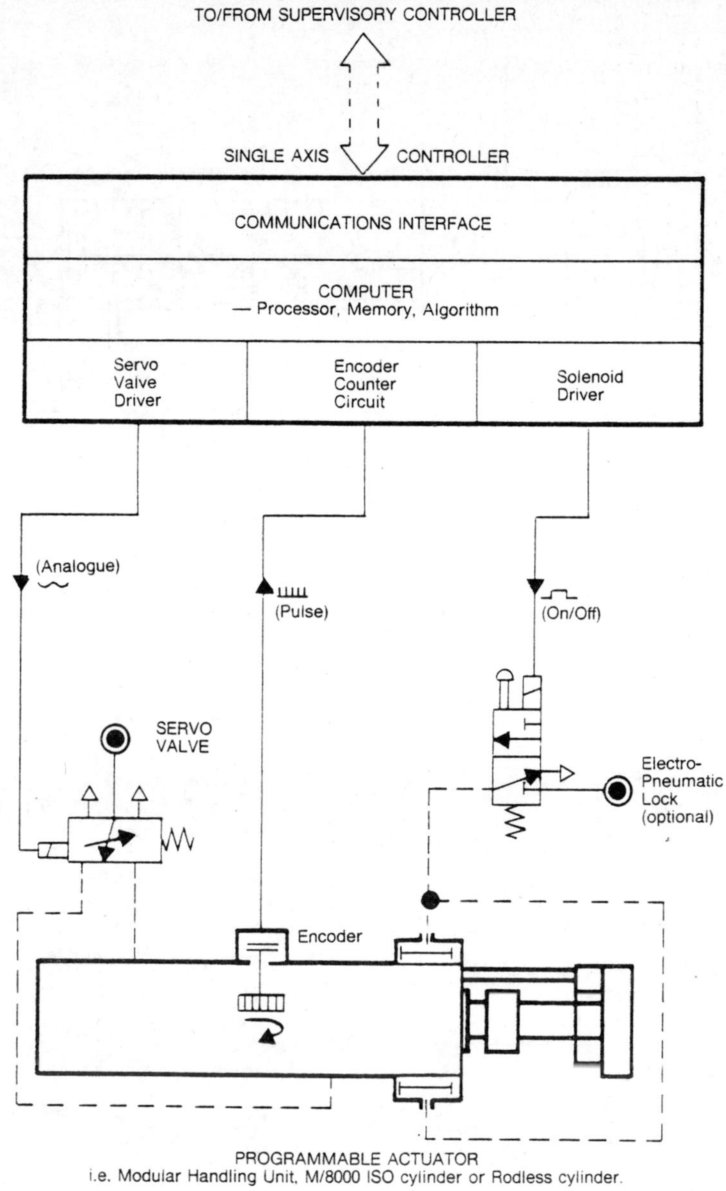

Fig 3 Servo control system

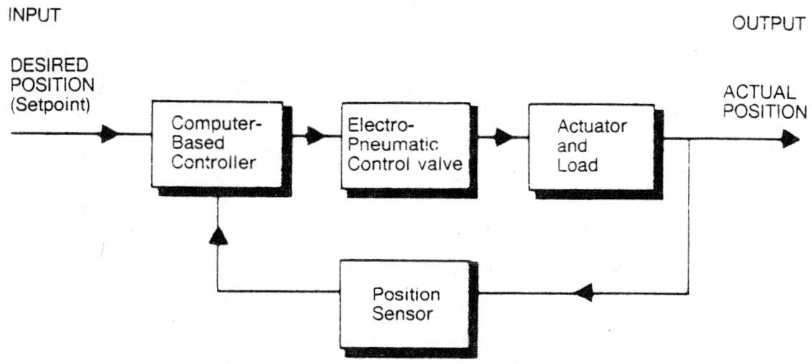

Fig 4 Block diagram of the servo system

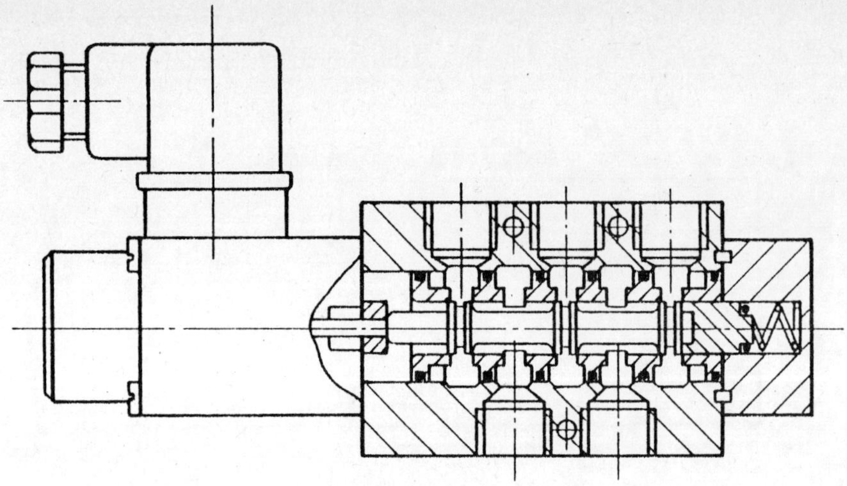

Fig 5 Servo valve

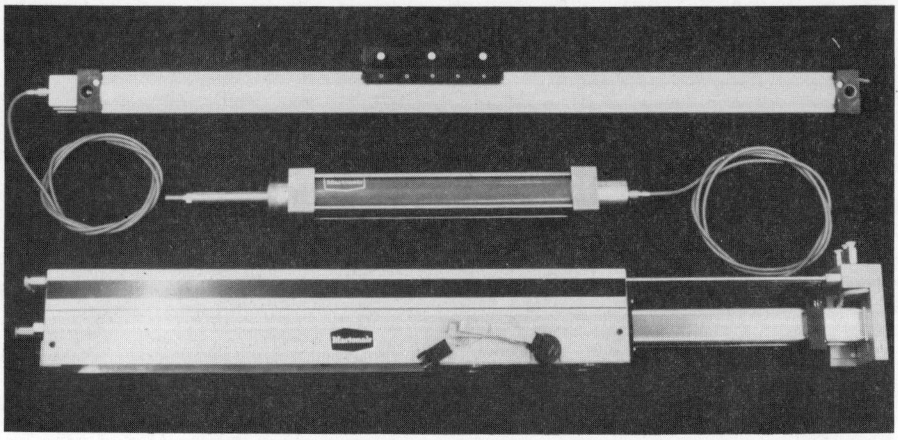

Fig 6 Actuators with feedback devices. Top, rodless cylinder; centre, conventional cylinder; bottom, handling module.

© IMechE 1985 C271/85

Fig 7 Palletiser

Fig 8 Loading equipment for use with a drilling machine

© IMechE 1985 C271/85

C266/85

Instrumentation in the high technology industries

C T BILLIET
Domnick Hunter Filters Limited, Birtley, Durham

SYNOPSIS Compressed air contamination levels are discussed and the quality standards reviewed. The paper outlines a new approach to compressed air purification recently introduced.

1. INTRODUCTION

Compressed air is known as the fourth utility and enjoys an excellent safety record. It does however have a reputation as being to a greater or lesser extent wet, oily and dirty. The electronics industry however has taken a harder look than most at quality requirements to meet the needs of an industry which can be affected more than most by contamination.

The various contaminants present in compressed air comprise atmospheric pollution including micro-organisms, compressor wear particles, bulk lubricating oil and aerosols, water vapour and condensed water, oil/water emulsion, rust, carbon and particles from pipework. Suitable treatment of the air involves the removal of the contaminants using aftercoolers, refrigerant or desiccant dryers and filters.

Coolers and dryers are used to remove water vapour and reduce the dew point of the air, thus preventing further condensation of water downstream. High efficiency coalescing filters are used to remove the other contaminants to extremely low remaining levels.

A review of possible contamination levels and compressed air quality standards reveals a somewhat haphazard approach. This is because users have not been able easily to quantify the amount of contamination present and are therefore unable to decide on acceptable levels for a particular use.

"Instrument air" is a much used term, but is not well defined and out of all air 'standards' is the one most likely to be contentious.

A new approach to air purification has already been introduced which takes a sound engineering approach to the design and manufacture of compressed air purification equipment. The new design takes full account of contamination present in the air and the requirements of users, particularly with regard to instrument air use. New test methods to verify contamination levels and air quality are also reviewed.

2. COMPRESSED AIR QUALITY

The recently published Pneurop Recommendation 6611/1984 Compressed Air For General Use Part 1 : Quality Classes (1), tackles the question of compressed air quality and puts the air into quality classes according to the intended application. It also suggests and recommends various test methods for the determination of air quality. It is doubtful however if the user of compressed air will be any the wiser for this since the mystery of compressed air quality still exists. How therefore can the industry remove the mystery and provide high quality air at a reasonable cost and in a reliable way?

It is not so long ago that a factory manager would purchase a compressor, after-cooler and receiver from separate sources. With the advent of packaged compressors the compressor system took a step forward. A neater package with built in aftercooler and sound insulated is now a common product. This action, welcome though it was, did not enhance the quality of the air being delivered, indeed in many cases contamination increased. There has been an increased use of refrig-erated and desiccant dryers complete with various types and qualities of filters in an attempt to clean the air to a satisfactory level.

3. CONTAMINANTS

Contamination can enter a compressed air system from three sources. These are namely, the atmosphere, the compressor and the piping system. Taking the atmosphere first; in a typical metropolitan environment there are something like 140 x 10^6 dirt particles per m^3 (4 x 10^6/cu.ft). Approximately 80 per cent of these are less than 2 micrometres in size (a µm is one millionth part of a metre or approximately 0.04 thousandths of an inch) and are not removed by the compressor intake filter. In addition to this, there are contaminants such as hydrocarbon vapours (from unburnt fuels and industrial processes) and water vapour. All of these are drawn into the compressor and their concentration increased by the compression process.

Inside the compressor, wear particles and lubricating oil are added to the air. The amount of oil contamination depends on the type of compressor and its condition but can typically vary between 5 and 50 mg/m³. To put this into context, a 280 L/s A.N.R.* (600 scfm) compressor carrying 30 mg/m³ (25 parts per million) of oil during an operational period of 6000 hours will deposit approximately 212 litres (47 gallons) of oil into the system. It is never an advantage to have oil from a compressor in the system as this oil is inevitably degraded and oxidized after the heat of compression. Usually it is acidic and can appear as a varnish like substance possessing properties completely opposed to lubrication. If synthetic oils are being used these can cause chemical attack on downstream components such as seals and polycarbonate parts. Carbonization of the lubricating oil in a reciprocating compressor is not uncommon, due to the temperature of compression, and the presence of 'hot spots' results in deposits of carbon which are again carried over. It is worth taking note at this point that even using a non-lubricated compressor does not necessarily generate 'oil free' air as quite large amounts of hydrocarbons can be drawn in from the atmosphere.

After compression the air then passes into the piping system. Normally, aftercooling reduces the amount of water vapour contained in the air but further cooling in the system causes more water to condense. Pipe-scale and rust are also present and when these contaminants mix together with the oil, carried over from the compressor, they form an abrasive, creamy sludge. This combination of contaminants causes costly problems with both the equipment and the product. These can be eliminated by suitably treating the compressed air using an effective filtration system.

Water vapour itself is sometimes a particular problem and cannot be tolerated in any amount, in which case a dryer is also necessary. When the air is discharged from an air cooled aftercooler at say 30°C and enters a compressed air system it will cool to the ambient temperature of say 20°C and as a result and assuming a pressure of 7 bar gauge (100 psig) about 1.6g/m³ will condense and present a real problem to all users of compressed air. In the earlier example of a 280 L/s A.N.R. (600 scfm) compressors over 6000 hours, more than 9680 litres (2130 gallons) of water will enter the system and must be removed.

4. CONTAMINATION LEVELS

The anticipated levels of various contaminants that may be expected to be present in a cubic metre of air at atmospheric pressure in an industrial environment are given in Table 1 and an investigation into available standards and specifications for instrument quality air covering the four main contaminants oil, water, dirt and vapour produced the results shown in Table 2. As can be seen the specified contamination levels vary considerably.

* Atmosphere Normale de Reference. Indicates that the rate of flow is free air at standard reference atmosphere conditions.

To determine what quality of air users should specifiy we have to consider what a particular system can tolerate, what equipment is providing satisfactory protection and what equipment is available.

It is academic to attempt to determine system tolerance to impurity levels since systems vary tremendously whilst at the same time there is equipment commercially available fulfilling the needs of users at acceptable cost. The biggest pollutant is however water in the form of condensate as not only can it combine with other contaminants to form abrasive sludges and sticky emulsion but it can also freeze in cold conditions preventing systems operating at all. So, it is reasonable to start by looking at water and while opinions differ as to the levels and size of dirt and oil permissible there is good agreement that air should be dried to -40°C pressure dew point. -40°C is used since the pressure dew point of the air must always be lower than the lowest possible ambient temperature through all seasons. API-RP550 (3) states pressure dew point must be at least 10K below the lowest ambient temperature in which any part of the equipment may be working. Refrigerant dryers are therefore only suitable where the ambient will always exceed approximately 15°C. A pressure dew point of -40°C can only be achieved commercially by the use of chemical dryers using adsorpotion material which in themselves require protection from bulk oil and water to prevent pollution of the adsorption surface, additionally particulates shed from the adsorption material exists which, while generally submicron in size, are in sufficient quantity to be unacceptable.

A picture begins to emerge

(a) Dryers need to be protected from oil/water in bulk and aerosol form by high efficiency coalescing filters and water separators respectively

(b) To prevent desiccant dust carry over entering the system submicron depth type or membrane filters are needed

(c) To prevent condensation of hydrocarbon vapours downstream activated carbon filters are required.

This approach can however result in a complex system expensive both in capital cost and running cost (Fig. 1). What is needed is a new approach to produce an integrated system containing all the above features.

5. TEST METHODS

5.1 Dewpoint

Dryness of compressed air is denoted by the Pressure Dewpoint, the temperature, referred to a specific pressure, at which water vapour begins to condense.

Pressure dewpoint is measured using specialised instruments manufactured by a variety of manufacturers. Psychrometer or electric hygrometers in which a change in

© IMechE 1985 C266/85

resistance or temperature of mirror surface is measured at the time frost formation begins or a piezoelectric hygrometer.

5.2 Oil Aerosols

Oil aerosols are present in very fine sub-micron form similar in size to tobacco smoke, (Fig 2) and have the ability to travel many kilometres down pipework systems without settling due to gravity and/or other effects. A quantitative method to determine oil carry over from high efficiency coalescing oil removal filters is specified in Pneurop Recommendation No. 6611/1984 Part 2 (1). Performance of oil removal filters was described by Billiet and Fielding in a paper presented at the 1984 International Compressor Engineering Conference. (4)

5.3 Oil Vapour

Oil vapour can be determined by a flame ionisation analyser and other types of specialised instrumentation. Adsorptive capacity of activated carbon can be determined by the chlorobenzene test method as generally described by Maggs and Smith (5)

5.4 Dirt

Dirt particle size can be determined using a cascade impactor, which can be operated at high pressure and temperatures. Numbers of dust particles can be determined using a particle counter or microscopic scanning in combination with retention on a membrane of suitable pore size.

Concentration can be determined by gravimetric methods which can be used also at high pressures whereas particle counters and light dispersion photometers can normally only be operated at atmospheric pressure.

Collecting efficiencies of filters can be predicted using microfibre filters as described by Stenhouse (6) and Billiet and Stenhouse (7).

6. A NEW APPROACH TO PURIFICATION

A completely new and innovative compressed air purification system has recently been introduced into the market (Fig. 3). It brings to the industry a new approach to design and manufacture of purification systems whilst using tried and tested methods to remove contaminants. Instead of a string of up to five stages of air treatment the PNEUDRI * has all these stages built into one modular system. This approach has the benefit of cost, space and energy savings, together with a high degree of engineering, ease of maintenance and reliability. The PNEUDRI brings top quality compressed air within the reach of all users and avoids the need to have so many different grades of air quality. Quite simply it makes selection of air quality easy by giving the best and provides users with cost effective air treatment that is affordable. Indeed it provides an air quality one cannot afford to be without.

* PNEUDRI and DRYFIL - trade marks of Domnick Hunter Filters Ltd.

Conventional pressure swing desiccant dryers are big and take up valuable floor space, they also in many cases have the appearance of a plumber's nightmare. They typically have many external pipes, valves, filters etc and are expensive to maintain.

By using a single extrusion to create a modular desiccant dryer (Fig. 4) the use of large welded steel tanks and the problems of welded pressure vessel approval are avoided, although the new dryer is designed in accordance with recognised pressure vessel codes.

Oil and water in bulk and aerosol form are removed by the integral high efficiency inlet filters. Water vapour is then adsorbed on the 3 mm dia. spherical pellets of a new type of activated alumina desiccant DRYFIL* contained in one half of the extruded columns. Heat is generated 3500 kJ/kg (1500 BTU/lb) of water adsorbed, and conducted through the integral aluminium module to the second desiccant tower simultaneously undergoing regeneration by purging with some of the dried air at lower pressure through the bed. Since desorption is endothermic the transferred heat from the tower on load helps to reduce the amount of purged air needed by 15 to 50 per cent.

The special construction of the dryer allows the columns to be filled with dessicant using the 'snow storm' filling technique developed by the Chemical Defence Establishment at Porton Down. This method gives optimum packing density and eliminates channelling.

Air is switched from one tower to the other by a simple full flow shuttle valve built into the base of the inlet filter housing (Fig. 5). The geometry of the adsorption columns has been optimised at a length to diameter ratio (L/D) of 13:1 (8) and since the design is modular increasing capacity does not mean giving up the optimum L/D ratio unlike conventional welded vessel systems. As a result of the new DRYFIL desiccant material and specialised filling techniques only approximately 50 per cent of desiccant by weight is needed compared to existing designs.

Optional activated carbon adsorption filters are located at the top of the modular columns with a design life of 8000 hours based on inlet oil vapour levels of 0.1 mg/m^3.

Simple full flow check valves are built into the top of the outlet filter housing where high efficiency depth filters are fitted. Alternatively a stainless steel outlet manifold (electro-polished) can be fitted together with 0.2 micron membrane filters pre-flushed to electronics grade quality requirements.

The control system in standard form is all electronic, controlling the operation of the functions and operating timed drain valves which remove collected liquids from the inlet filters. All valve controls are pneumatically operated by printed circuit board mounted solenoid valves and are contained in a control box. All pneumatic controllers are also available for flame-proof environments.

A heat regenerative version, requiring a significantly lower purge air requirement due to the assistance of heat energy to effect desorption, is also available. Utilizing positive temperature co-efficient semi conductor ceramic heating elements, positioned in the top distribution manifold and thus not in physical contact with the desiccant, their 'honeycomb' structure provides maximum heat dissipation to the purge air at minimal pressure drop. Being intrinsically safe, even under no flow conditions these PTC heaters self regulate their applied thermal load, allowing extremely efficient operation, and enable regenerative cycle times to be reduced to a minimum.

6.1 Dryer Specification

This modular construction allows additional pairs of columns to be added up to a maximum of eight pairs in a single bank and these can then be double, triple or quadruple banked to provide drying capacity at 7 bar from 35 L/s ANR up to 1100 L/s ANR (Fig. 6). Larger or smaller models can easily be produced.

The resulting compressed air has the following maximum contamination levels

Dewpoint	-40°C
Dirt	0.01 µm
Oil aerosol	0.01 mg/m³
Oil vapour	0.003 mg/m³

This specification meets or exceeds the best of all standards for any contaminants as shown at the foot of Table 2.

7. FEATURES OF PNEUDRI

The particular features apart from its performance specification and absence of exposed pipework can be summarised as follows:-

Full instrumentation as standard equipment with gauges for air inlet and air outlet pressure for both towers. Indicator lights for power, drying, purging, moisture indicator for outlet air (dew point meter optional.)

Custom - programmed microprocessor control enclosed in control box Class IP55 to BS5490 (9) (Pneumatic controller optional). All electrical wiring and connections contained in control box.

Functionally stylised cabinet readily removable for servicing of dryer components.

Purge silencers and sound insulation for quiet operation.

Simpler, more reliable purge valves with large opening for maximum efficiency.

Automatic timed drains at bottom of inlet manifold for removal of condensed water and oil.

Versatile mounting system adapts dryer for floor, wall or overhead mounting.

Desiccant bed has optimum length-to-diameter ratio. Modular construction allows this ratio to remain constant for all dryer sizes.

Integral inlet shuttle valve designed for high reliability and low pressure drop.

Modular construction from aluminium extrusion and plate is fully certified. Components alochromed and externally epoxy powder painted for resistance to corrosion and abrasion. Bolt-together construction provides easy access for servicing.

Built-in inlet coalescing filters with automatic timed drains.

Heat transfer from adsorption to desorption column provides greater operating efficiency.

Built-in outlet filtration (stainless steel outlet manifold and membrane filter, optional extra)

Built-in oil vapour/odour removal (optional extra).

Integral check valves designed for high reliability and low pressur drop.

Purge orifice infinitely adjustable for precise purge flow adjustment (pre-set at factory).

Compact design with integral components provides exceptional space efficiency.

5 year service contract available.

8. SUMMARY

A suitable specification for the quality of compressed air for instrumentation especially in the high technology industries has been developed and suitable equipment to meet this specification produced.

Air of the required quality can be provided by equipment that has been available for many years when assembled as shown in Fig. 1 but such an arrangement is complex, expensive and sometimes lacking in long term reliability.

The overall performance and compactness of this new development in compressed air processing exceeds anything previously available.

Table 1 Level of contaminants to be expected

Contaminant	Source	Typical Concentration
Dirt particles	Atmosphere	up to 140 x 10⁶/m³
Carbon	Burnt Oil	up to 10 mg/m³
Water	Atmosphere	up to 11 g/m³
Rust	Pipework	up to 4 mg/m³
Oil	Compressor lubricant	5 - 50 mg/m³
Oil/water emulsion	Mixture of oil & water	up to 11 g/m³
Vapour	Gaseous oil	0.05 - 0.5 mg/m³
Micro-organisms	Atmosphere	up to 3850/m³ (2)
Unburnt hydrocarbon	Atmosphere	up to 0.5 mg/m³

© IMechE 1985 C266/85

Table 2 Review of international standards on contamination

ISSUING AUTHORITY & TITLE	REF NO	ORIGIN	CONTAMINATION LEVELS SPECIFIED: SOLIDS	WATER	WATER VAPOUR	OIL MIST	OIL VAPOUR
American Petroleum Institute "Air Supply Systems"	RP550 Sect. 9	USA 1980	<5 µm	Nil	Pressure Dewpoint 10 K below ambient	Nil	Nil
AmericanNational Standards Inst. "Quality Standards for Instrument Air"	ISAS7.3	USA 1981	<3 µm	Nil	Pressure Dewpoint 10 k below ambient	1 mg/m^3	Not Specified
Ministry of Defence "Compressed Breathing Air For Aircraft Systems"	16-8 Issue 2	UK 1972	<10 µm	Nil	<7 mg/m^3 atmospheric	5 mg/m^3	Not Specified
International Standards Organisation. "Compressed non breathing air for use in aircraft"	IS02434	ISO 1973	<10 µm	Nil	<20 mg/m^3 atmospheric	<5 mg/m^3	Not Specified
GOST "Industrial Purity-Grades Of Contamination For Compressed Air"	17433	USSR 1972	0.2 µm up to >100 µm	Nil up to 10g/m^3	Not specified	Nil up to <100 mg/m^3	<0.3 mg.m^3
PNEUROP Compressed Air For General Use "Quality Classes" Process Control Instruments	6611 Part 1	Europe 1984	1 µ (1mg/m^3)	Nil	10°k below ambient or -20°C	1 mg/m^3	
Clean Dry Air (C.D.A.) Specification U.S.A.	Un-official	USA	0.2 µ	Nil	-40°C	<0.5 mg/m^3	0.5 mg/m^3
Best of all above grades			0.2 µ	Nil	-40°C	<1 mg/m^3	
Domnick Hunter Filters Limited	PNEUDRI *	UK 1985	down to 0.01 µ	nil	-40°C or better	0.01 mg/m^3	0.003 mg/m^3

* Including Activated Carbon Stage

REFERENCES

(1) PNEUROP Recommendation No 6611-1984 Compressed Air For General Use
Part I : Quality Classes
Part 2 : Testing
British Compressed Air Society 1984

(2) BOURDILLON, R.B., Lidwell, O.M., and Thomas, J.C. J. Hyg. Camb., 1941, 41 197

(3) API Recommended Practice 550, Manual on Installation of Refinery Instruments and Control Systems. Part I Process Instrumentation and Control Section 9 - Air Supply Systems. American Petroleum Institute 1980

(4) BILLIET C.T. and FIELDING R.M. A Quantative Method To Determine Oil Carry Over From High Efficiency Coalescing Filters in Compressed Gases Under Varying Dynamic Conditions and the Interpretation of Results to Assist Filter System Design. International Compressor Engineering Conference Purdue University USA July 1984

(5) MAGGS F.A.P. and SMITH M.E. A Non-Destructive Production Line Test For Vapour Filters, Filtration & Separation, 1973, July/August, 347-356

(6) J.I.T. STENHOUSE, Mechanisms of Gas Filtration, Filtration & Separation 1975, 70-76

(7) BILLIET C.T. and J.I.T. STENHOUSE. A Theoretical Study Of Filter Systems Used In The Production Of Sterile Air - 1985

(8) J.P. AUSIKAITIS. Pneumatic System Air Drying by Pressure Swing Adsorption, American Chemical Society 1977

(9) BS5490 Specification for - Classification of degrees of protection provided by enclosures. This is identical to IEC 529 : 1976, British Standards Institution 1977

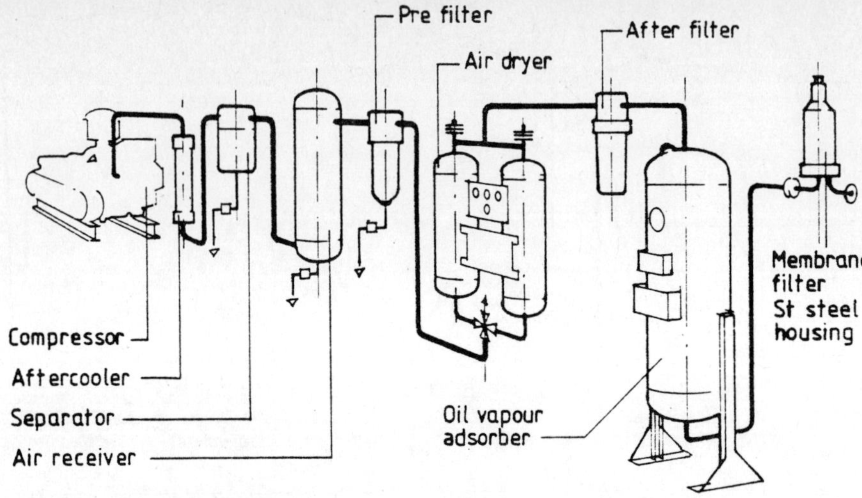

Fig 1 Typical current system

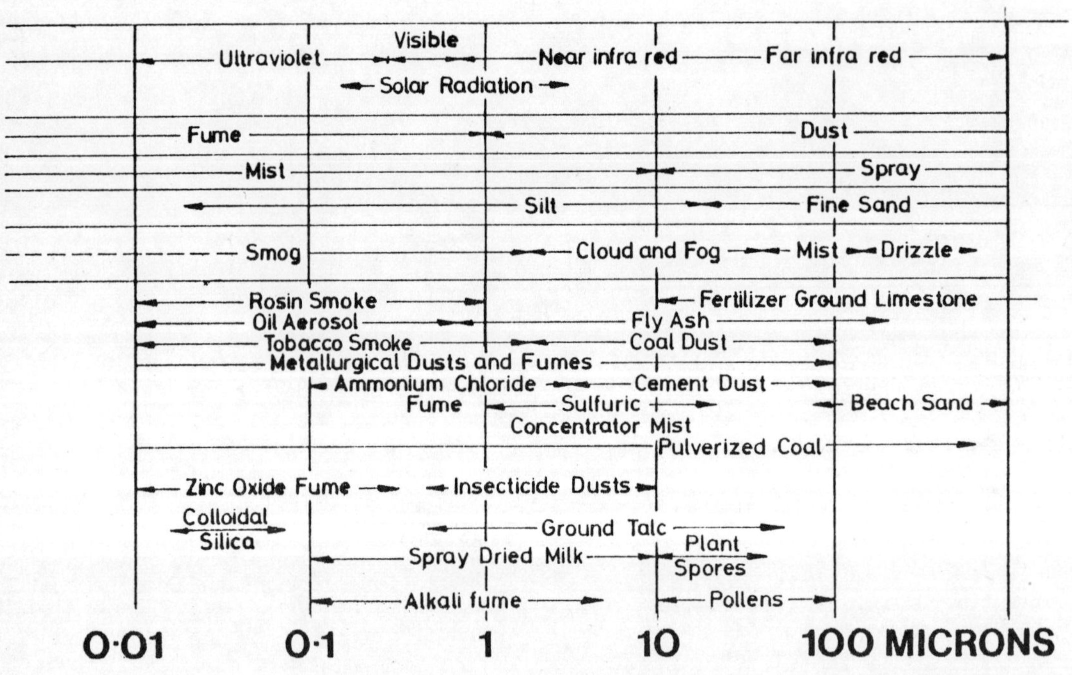

Fig 2 Typical particle diameters μm

© IMechE 1985 C266/85

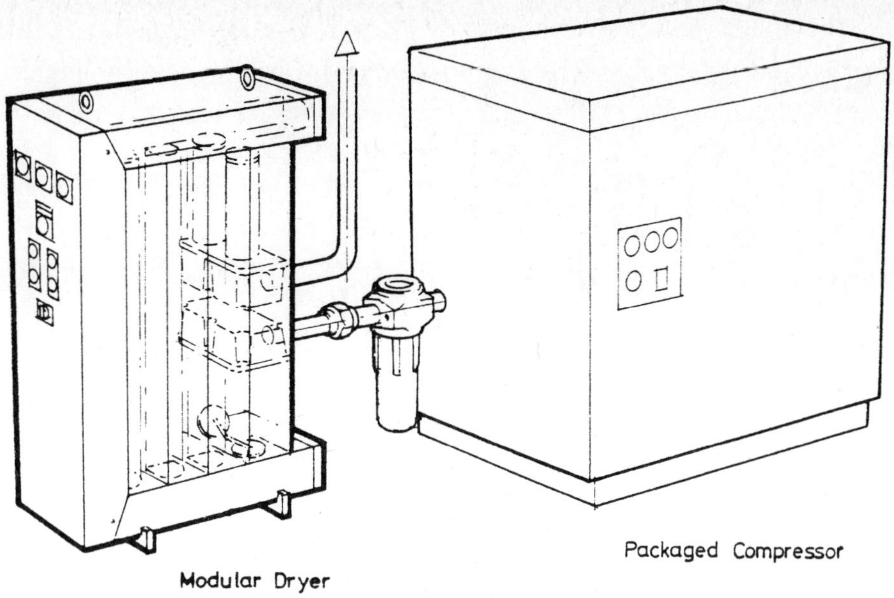

Fig 3 Compressor dryer assembly

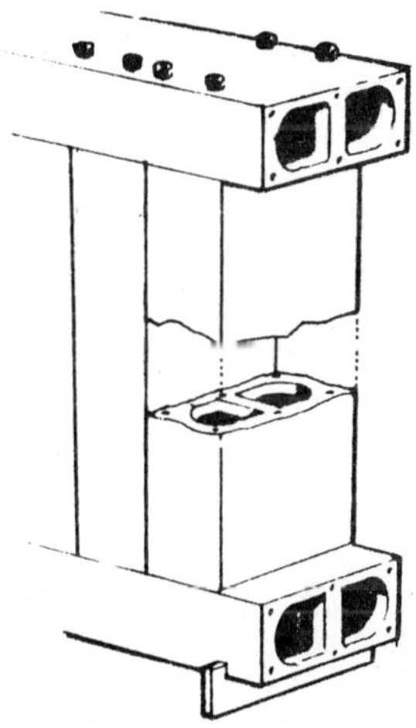

Fig 4 Basic extrusion for modular dryer

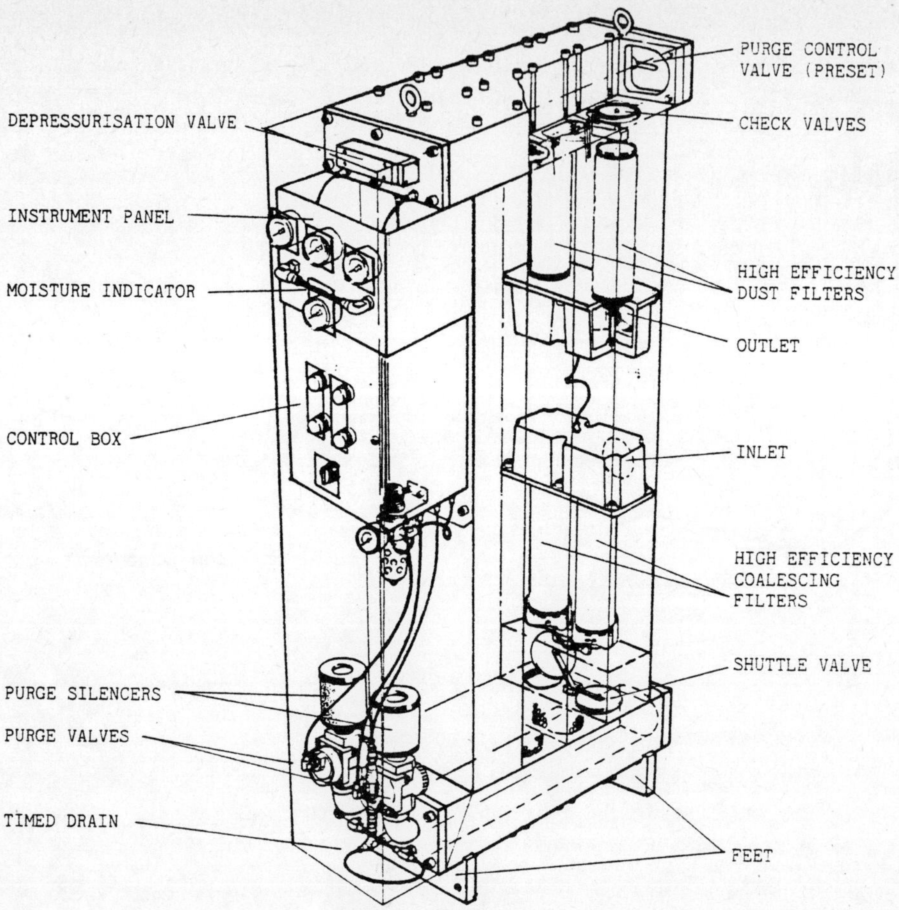

Fig 5 Arrangement of basic dryer assembly

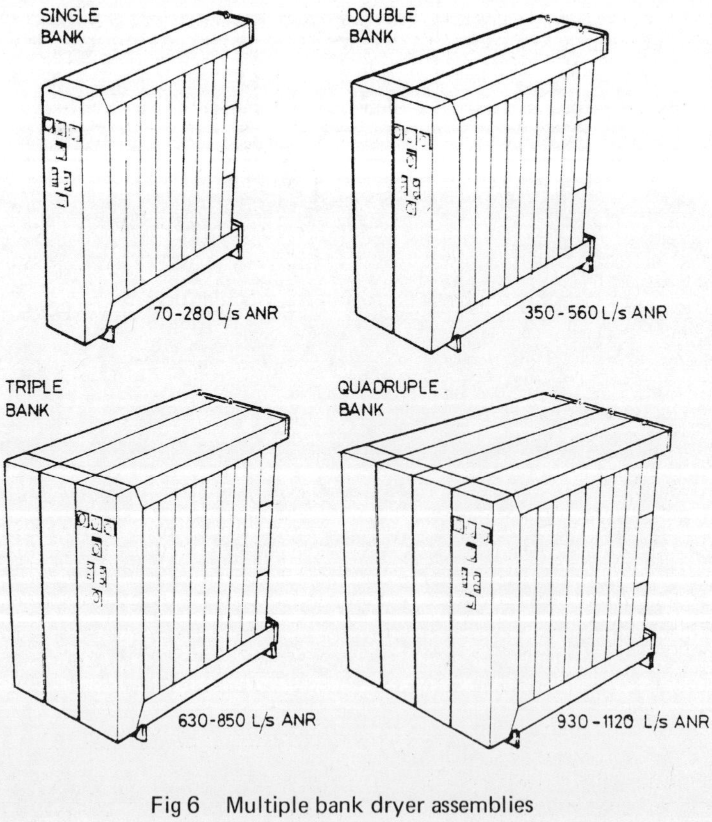

Fig 6 Multiple bank dryer assemblies

© IMechE 1985 C266/85

C265/85

Trends in prime mover choice for powered hand tools

D S GILCHRIST, AMIMechE, **R J OATHAM**, CEng, MIEE and
DR C R SIMS, CEng, MIEE
PA Technology Melbourn, Royston, Hertfordshire

SYNOPSIS Many of the traditional choices of prime movers for powered hand tools are being re-examined in the light of technological advances. In particular, pneumatic tools, for a long time the almost universal choice for high powered industrial tools are being replaced by new generations of electric drives which combine long life owing to brushless construction with superior control. This trend has been accelerated by the advent of new permanent magnet materials which have permitted much higher power/weight ratios to be realised.

Hydraulic hand tools are found in a few high power applications, however in the majority of lower powered tools, electric or pneumatic drives are preferred and this situation is unlikely to change.

The paper compares the performance of the range of tools and looks at how they might develop in the future. Figures are provided for power and torque against weight and volume, together with typical parameters of achievable control, e.g. speed range, regulation.

1 INTRODUCTION

Powered hand tools are found in most sectors of manufacturing and construction industries as well as in medical and domestic applications. The choice of motor for these tools has been determined principally from power source availability, but the nature of the application often dictates a selection on the basis of other factors such as intrinsic safety or power to weight ratio.

There are four categories of prime mover in common use, for powered hand tools:

Electric
Pneumatic
Hydraulic
Gasoline

The fundamental characteristics of the prime mover often determine the performance of the tool and may not match all the requirements of the application. In some cases, principally electric motors, the capability to tailor the fundamental characteristics to suit the application more closely is leading manufacturers to reconsider their traditional prime mover choices.

Advances in electric motor technology; brushless construction and electronic control, are challenging the position of the Universal Motor within existing electrically powered tools. Furthermore, the new electric drives may provide a viable alternative to pneumatic and perhaps hydraulic motors in many applications. This paper reviews powered hand tool applications, compares the characteristics of Electric, Pneumatic and Hydraulic motors and identifies how the traditional prime mover choice may be affected by future developments.

2 POWERED HAND TOOLS

2.1 Scope

Within the terms of this paper, 'Powered Hand Tools' include the following categories:

(a) Drills, Reamers
(b) Grinders and Sanders
(c) Circular, Reciprocating and Chain Saws
(d) Breakers
(e) 'Others' - Nut runners, screwdrivers, nailers, riveters, band strappers, etc.

Within each application category, the principal determining factors are:

Power Source Availability
Portability
Performance Characteristics and Duty Cycle
Environmental and Safety
Durability
Quality and Reliability
Cost
Ergonomics: Size, weight, 'Umbilical' Requirements
Control
Efficiency

Nonetheless, customer preference in many applications is highly defined ; being based on previous experience and an unwillingness to proliferate tool types. Any changes in prime mover specification, for example in the construction industry, will be slow and only on the basis of well defined advantages at negligible risk.

2.2 Market Segmentation

The breakdown, in £ value terms, of the powered hand tool market is estimated to be divided between Pneumatic, Hydraulic, Electric and 'Others' (Gasoline, etc.) as follows:

This data is shown graphically in Figure 1. (Source : PA, market survey)

By industrial sector (% of sector):-

	PNEUMATIC	HYDRAULIC	ELECTRIC	OTHERS
Domestic/Mobile Service	-	-	100	-
Medium Industrial, Assembly	35	-	65	-
Civil Engineering/Lumber	68	10	6	16

and by application (% of total):-

	PNEUMATIC	HYDRAULIC	ELECTRIC	OTHER	ALL TYPES
Breakers	5	3	-	-	8
Drills	16	1	28	1	46
Sanders/Grinders	5	-	13	-	18
Saws	2	-	4	5	11
Others	10	1	5	1	17
TOTALS	38	5	50	7	100

2.3 Market Opportunities

The market breakdown given above indicates the extent of any potential shift in prime mover choice. Nevertheless, customer preference and power source availability will be limiting factors. For example, the traditional dominance of pneumatics in construction work will be likely to prevail for many years to come.

However, a significant opportunity for substitution of pneumatic by electric motor lies in the medium Industrial and Assembly sector.

Examination of the breakdown by application suggests that substitution of pneumatics will be necessary for electric motors to take significantly larger market share; Electric and Pneumatic tools together constitute some 88% of the total powered hand tool market.

The interdependence of sales volume and cost is a major factor in the successful expansion of electric power tools. For example, typical 200 watt drill universal motors are produced in millions per annum on dedicated, automated manufacturing facilities. The finished manufactured cost may be only as much as 25 - 30% greater than the raw material cost.

3. COMPARISON OF PRIME MOVER CHARACTERISTICS

Figures 2. through 5. show Power and Torque versus Weight and Volume for the three motor types. The figures can be summarised as follows:

	PNEUMATIC	HYDRAULIC	ELECTRIC PEAK	ELECTRIC CONTINUOUS
Sample ranges:				
Power Watt	63 - 720	750 - 11000	30 - 2000	20 - 1000
Torque NM	.08 - 7.6	0.4 - 42	.07 - 10	.02 - 2.2
Specific Power kW/kg				
Mean:	0.3	2.0	0.5	0.2
Best:	0.72	5.5	1.1	0.5
N/L Speed r/min	1300 - 24000	600 - 30000	2800 - 40000	
Weight kg	0.3 - 4.5*	0.5 - 4.5*	0.2 - 4.5*	

*The ranges of Power and Torque for Pneumatic and Hydraulic motors corresponds to the selected weight limit of 4.5 kg. This figure has been chosen somewhat arbitrarily to represent an acceptable upper weight limit of a motor for a powered hand tool.

Higher Powers and Torques are of course available in each category in specific instances outside the selected weight ranges.

4. PNEUMATIC MOTORS

The potential vulnerability of the market position of pneumatic motors has been indicated above, however it is most likely that manufacturers will respond with improvements and feature enhancements. Within the spectrum of currently available motors there is a wide range of specific outputs particularly in the 100 - 300 Watt range. Furthermore, some of this variation can be attributed to efficiency differences. Given the attention devoted in recent years to reducing energy costs by improving compressor efficiency, it is perhaps surprising that motor efficiency and appropriate sizing are not given greater emphasis. To illustrate this point, eight different motor designs in the range 160 - 300 W have efficiencies ranging from 17 to 32%:

MOTOR	TYPE	POWER W	POWER/WT W/Kg	ISENTROPIC EFFICIENCY
1	Small vane reversible	163	285	17%
2	Small vane unidirectional	200	180	18%
3	Small vane reversible	78	140	20%
4	Small vane unidirectional	190	330	21%
5	Large vane unidirectional	267	300	26%
6	Small vane unidirectional	200	400	31%
7	Large radial piston	283	80	31%
8	Small vane unidirectional	200	525	32%

The last motor in the table above is a recently developed air motor for medical applications. (Where the intrinsic safety of air tools in a flammable anaesthetic environment is crucial). This particular motor has nearly twice the power to weight ratio of average air motors and a high efficiency, consuming 50% of the airflow requirements of the worst motors. These figures represent what is probably currently achievable in terms of pneumatic hand power tools in this power range. It should be noted that continuously rated electric motors are comparable to air motors in terms of average and best specific power and that for intermittent duties, the electric motors are somewhat better.

Control Aspects

Torque and speed control have been widely available in air tools for a number of years. However, recent developments in bolt tightening have set standards of torque control using electronic sensing which are unlikely to be met by traditional mechanical means. Future developments will therefore favour electric motors for ease of control in these applications. Speed control of pneumatic motors has usually been in the form of speed limiting rather than all-speed governing. It is possible that should the need arise mechanical/pneumatic all-speed control could be provided though at a significant cost penalty. Again the relative ease of such control in certain electric motor drives is highly attractive.

© IMechE 1985 C265/85

5. HYDRAULIC MOTORS

Hydraulically powered hand tools are still a comparative rarity except for the case of high power breakers where the exceptional specific power of hydraulic motors can be exploited. Indeed, for powers above 2.5 kW, hydraulic motors are the only reasonable choice given the weight constraints of a powered hand tool. At lower outputs, hydraulic motors do not offer clear advantages over the pneumatic or electric alternatives. There is considerable customer resistance due to the requirement for a power pack, relatively inflexible hoses and, not least, the traditional problem of leakage.

To summarise, the number of very high power hand tool applications is limited and it is therefore unlikely that hydraulic motors will find such wider market acceptance.

6. ELECTRIC MOTORS

Thus far, the discussion has for reasons of simplification referred to 'Electric Motors' as a single category - this becomes increasingly difficult since recent developments are expanding both the types of motor available and the characteristics of each. The majority of electric hand tools have been powered by brushed Universal Motors operated from the mains supply. As market pressure mounts within the traditionally electrically powered tool sector and as opportunities to replace non-electric prime movers are found, novel motor configurations are being developed helped along by advances in semiconductor and magnet technology. In many cases, the ability to tailor the motor to the application is crucial to the optimisation of performance and acceptability of the final product.

The following discussion describes the two basic families of hand tool motors: brushed and brushless.

6.1 Brushed or Commutator Motors

There are three types of electric motor within this category:

- Wound field, series connected, AC or DC - The Universal Motor
- Permanent Magnet field, DC only - The PM Motor
- Wound field, shunt connected, DC only - The Shunt Motor

Since the majority of hand tools are mains operated, Universal Motors are most widely used, however low voltage battery tools tend to use PM motors.

6.1.1 Performance Characteristics

The Universal Motor has a high power to weight ratio compared with other mains frequency motors because of its high operating speed. The no-load speed is only limited by friction and windage and the motor has a high stall or starting torque. A typical speed/torque characteristic is shown in Figure 6. Speed range control without power loss is generally achieved by a single or multi-speed gearbox. Speed control is also possible by voltage control (usually via diode or S.C.R.) but unless electronic feedback control is used, regulation is poor and power is reduced.

The PM motor has a linear speed/torque characteristic again with a high stall torque but with a definite, limit to no-load speed. (Figure 6.) The overall speed range is somewhat less than that of the Universal Motor.

The Shunt Motor has a similar speed/torque characteristic to that of the PM motor but for the demagnetising effect of armature reaction, which reduces stall torque. Variable speed control can be readily achieved by varying the field current or armature voltage independently.

7.1.2 Motor Developments

Universal Motors have evolved considerably, with developments such as integration of the motor within the tool housing and increased power to weight ratios achieved by increasing motor speeds (and therefore using higher gear ratios) and to a certain extent increasing motor temperatures.

Improved reliability and user tolerance has been brought about by the following factors:

(a) Improved Materials

- Enamelled wire coatings with higher temperature capability, better toughness and quality.
- Impregnation varnishes - higher strength epoxy and polyester resins allow armatures to run hotter and at higher speeds.

  Slot Insulation synthetics such as polyester replace paper and 'leatheroid' thereby improving mechanical and thermal properties.

(b) Improved Processing

- Hot staking of commutator connections to replace solder.
- 'Trickle' impregnation to replace 'dip and bake' methods.
- Automatic assembly - particularly automatic winding and commutator staking - give consistently higher quality results at lower cost.

The use of plastics for tool bodies (providing Double Insulation) presents more scope in providing a lighter, safer and cheaper tool. In particular, the development of the Integral tool designed specifically for the task has been highly successful allowing optimisation of motor characteristics.

Electronic Control was first introduced as a crude speed control using diode and thyristor switches. Nowadays, motors can have speed and current sensing to give a wide range of speed and torque control. Microprocessor developments will offer enhanced features such as optimising speed/torque for a specific load which may be selected by the user. It may well be expected that mechanical overload clutches found in screwdrills could be replaced by appropriate electronic control.

Increasingly however, some of these developments are highlighting the limitations, particularly durability, of the commutator and brush gear and attention is becoming focussed on the brushless motor as an alternative.

7.2 Brushless Motors

The term 'brushless' includes the following families of motor:

- DC Permanent Magnet Motors
- Reluctance Motors
- Induction Motors
- Hysteresis Motors
- Stepper Motors

These motor types are finding increasing usage due primarily to advances in semiconductor technology: Power Devices are being developed to meet higher output requirements and hence larger motors. Improved performance characteristics in switching and back emf capability are to be found at all power ratings.

Small Signal Devices are available to control speed (tacho sensors) and load, sense motor position and provide current limit protection.

There can be as many possible motor configurations as applications, for example, a DC motor specification could include the following:

- Construction	- Internal/External Rotor, Framed/Frameless, Pancake, Hollow Shaft.
- Windings	- Slot configuration, airgap, printed or wire wound.
- Drives	- Sinusoidal, Stepped (e.g. 6 step inverter). PWM (including sine weighted PWM).
- Switching	- Bipolar, GTO, Darlington, SCR, Power Mosfet.
- Control	- Load Angle, Conduction Period, Current Limits - Microprocessor/Hard Logic.

This flexibility in design implies that there are not 'off-the-shelf' solutions for general usage. Each application has to be matched to a specific design in order to optimise the final tool performance characteristics. These might include;

- Improved torque/speed capability
- Enhanced Durability
- Control Sophistication

The superior commutation achievable with solid state switching permits operation at higher speeds, up to and beyond 80,000 r/min, this in turn leads to higher power to weight ratios.

Some indication of what is achievable is demonstrated by a recently developed brushless motor for a medical drill, incorporating rare earth magnets:

Power Watt	110 (Peak)	35 (Continuous)
Torque NM	0.1 (Peak)	0.02 (Continuous)
No-load Speed r/min	57,000	
Maximum Power Speed r/min	20,000	
Weight kg	0.12	
Volume CC	26	
Specific Power kW/kg	1.0 (Peak)	0.3 (Continuous)
Ref: Average Electric Motor	0.5 (Peak)	0.2 (Continuous)
Power Density W/cc	4.2 (Peak)	1.4 (Continuous)
Ref: Average Electric Motor	1.3 (Peak	0.5 (Continuous)

This drill was developed to operate in an intermittent mode and achieves over three times the power density (Power/Volume) and twice the specific power (Power/Weight) of conventional electric power tool motors (N.B. the weights and volumes do not include the control electronics which were remote to the tool in this particular case). In another brushless motor development, an indication of the potential control sophistication is illustrated by accurate speed control achievable throughout the range from 800 to 80,000 r/min.

For hand tools the most appropriate types in current applications are either PM (DC) or reluctance motors. As a replacement for a Universal Motor, the Reluctance Motor is cost competitive and offers significant durability advantages in certain application. For special purpose applications, perhaps currently powered by pneumatic motors, the PM motor is a potential candidate with a high degree of versatility.

7. CONCLUSIONS

Of the prime mover types considered, Pneumatic Hydraulic and Electric, the majority of hand tool motors are either Electric or Pneumatic. Hydraulic hand tools are effective only at very high power outputs where neither Pneumatic nor Electric motors are available at acceptable weights. It is, therefore unlikely that hydraulically powered tools will expand much beyond their existing market share.

Electric tools dominate the domestic and DIY markets and will probably remain unchallenged. The electronically controlled Universal Motor will continue to be developed in these applications, but will meet increased competition from brushless motors.

The large market share of pneumatic tools in industrial applications is unlikely to expand and will come under increasing pressure from novel electrically powered tools exploiting advances in motor technology.

The unique ability of the electric motor designs to be closely tailored to the tool application will imply a proliferation of motor types and ensure increasing competitiveness of electric power tools.

© IMechE 1985 C265/85

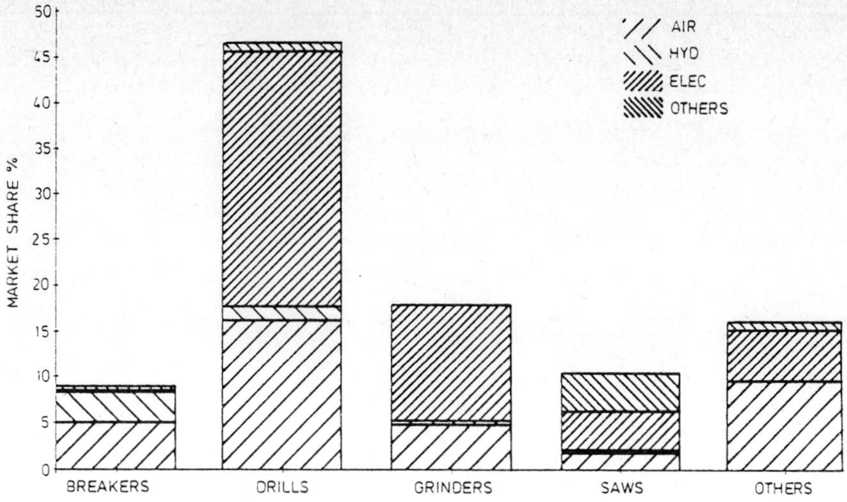

Fig 1 Hand tool market structure

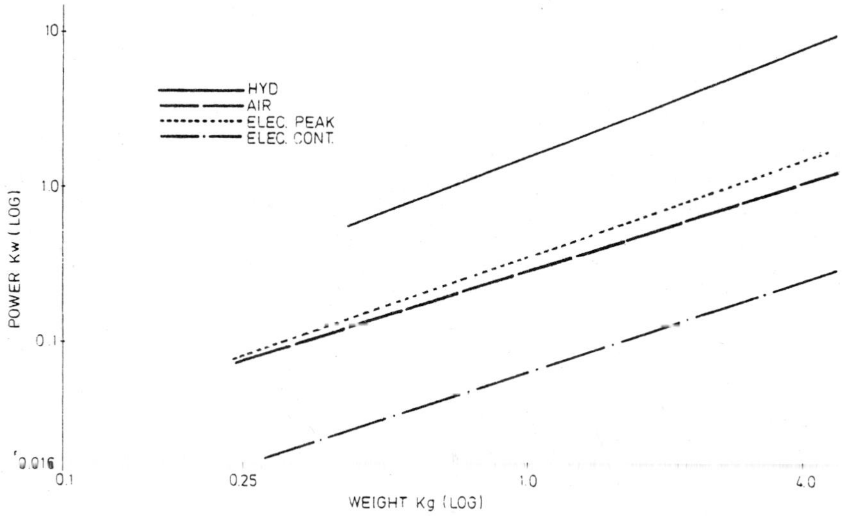

Fig 2 Power versus weight

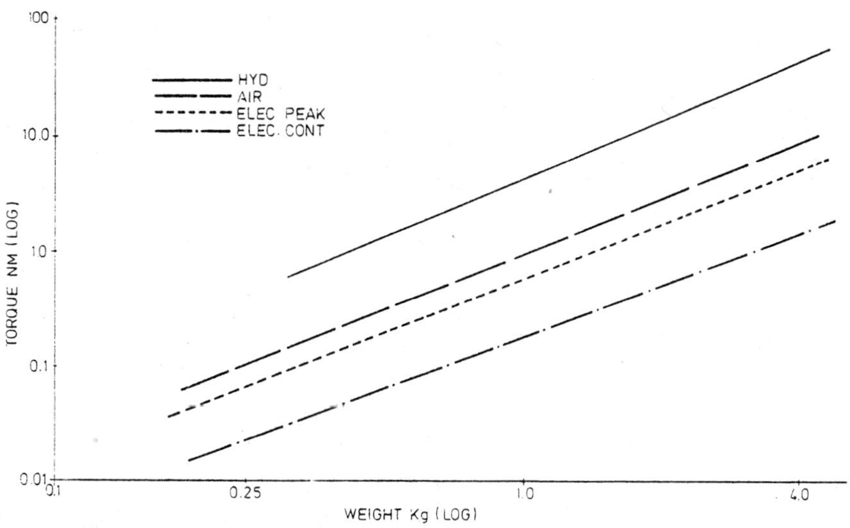

Fig 3 Torque versus weight

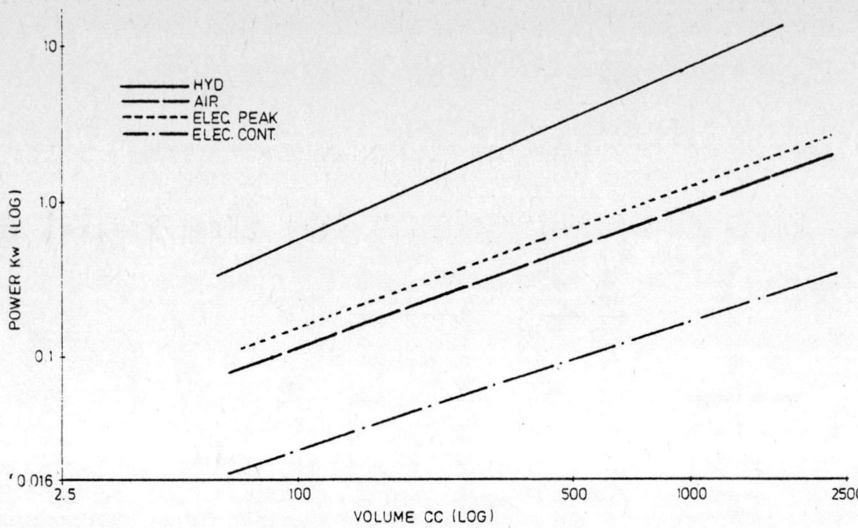

Fig 4 Power versus volume

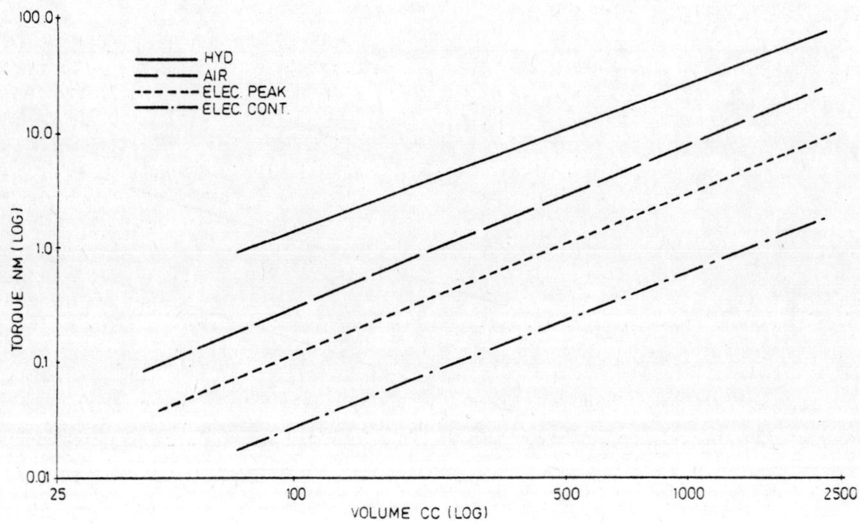

Fig 5 Torque versus volume

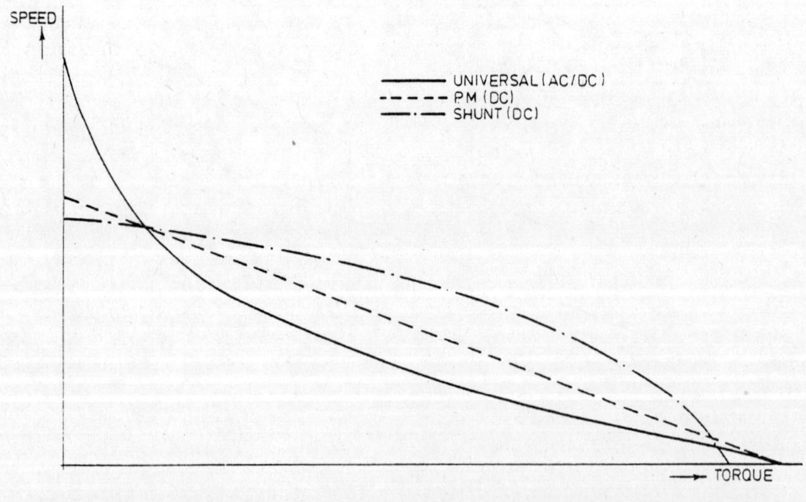

Fig 6 Typical speed/torque characteristics – commutator motors

© IMechE 1985 C265/85

C277/85

Choosing the drive system for the right application

I S LARMAN
Island Associates, Tadcaster, Nottinghamshire

SYNOPSIS.

This paper covers the basic control parameters of electric, hydraulic and pneumatic drives. It has been limited to variable speed drives. Consideration is given to AC motor controllers, hydraulic piston motors and air motors. The operating efficiency, torque and speed of a given drive are discussed and comparisons drawn. The advantages of one drive over another are shown as are their relative costs. the possibilities of a drive being controlled by microprocessors are also considered.

1. LIMITATIONS OF POINTS FOR CONSIDERATION

For constant speed applications the choice of drive is obvious. An electric motor and gearbox will provide all the parameters that are needed. However, if a variable speed is required then the choice is not always so simple.

When discussing which drive to use for a given application, personal preference, familiarity and commercial attitudes automatically come to the fore. This paper could extend to all types of drive system, encompassing many varied applications. However, consideration has been limited to variable speed drives in the 3-45 Kw range.

There is little choice for the practising engineer for variable speed drives above 45Kw, when the drive must be powered by hydraulics.

The engineer frequently requires a variable speed and torque output and, therefore, consideration of these parameters must dictate our approach and application.

2 VARIABLE SPEED DRIVES

What is meant by variable speed drives is that during a machine or process cycle the output speed must be changed to meet the different parameters. For example: the diameter of paper or cloth being wound on to a reel will increase so that the drive speed of the reel must change in relationship to the length being applied, assuming that the winding speed is constant.

2.1 Electric Drives

Electric motors which are capable of giving a variable output between 20% and 200% of their rated performance are usually controlled by an electronic device which changes the phases of alternating current. Gear boxes are used in very large systems to enable high power to be converted to a variable output, but this paper has chosen to exclude gear assisted drives.

Variable speed electrical control systems incorporate a means of switching the input current of the mains. This is achieved by either increasing or decreasing the frequency of the current at a ratio equal to the increase or decrease of speed which is required.

A typical block diagram is shown (fig 1) highlighting the microprocessor control of the motor speed. Inputs are provided through a keyboard to give information relating to Voltage, frequency, acceleration time, deceleration time, set frequency, minimum frequency, and maximum frequency, start, stop, brake, reverse increase speed and decrease speed are also set. The outputs are given as a simple display message of the analogue values of Voltage and frequency speed and status of the drive.

Pulse width modulation is generated by the microprocessor at a maximum frequency of 100Hz. Pulsing the gate of a thyrister at this frequency will double the speed of the electric motor. Obviously there will occur a drop-off in torque from the motor when it is rotating at speeds which are above the base speed of a motor. The power rating of the frame speed must be maintained.

The graphs shown (fig 2) give details of the percentage of the torque characteristic of the motor in relationship to the percentage of base speed. It is obvious that the torque reduces to a half when the speed of rotation reaches 200% of the base speed. However, some motors respond differently and will probably give an output of only 25% torque at maximum speed.

Loss of torque is also experienced when the motor is rotating at less than the base speed, for example: when the motor is rotating at 150 rpm (1/10th the base speed) the output torque is 75% of full load torque. This drop-off can be compensated for by boosting the low frequency voltage and in time increasing the torque.

2.2 Pneumatic Drives

Pneumatic drives are used in many different environments and will operate in conditions which will normally preclude the other methods of rotation. It is common to see, even in small factories, a ring main of compressed air. Adequate connecting points are readily available allowing quick and clean installation.

Air motors, the colloquial term used for pneumatic drives, are manufactured with either pistons or vanes as part of their construction. They can be small, used for self-feed drills, or large and robust, suitable for winch-drum applications.

Vane type air motors are used primarily for high speed applications such as self-feed drills. The dentist's drill is well known for its very high speed and as such reduces the discomfort that the slower belt driven drills used to have.

The maximum drive power of an air motor is about 25KW. The piston motor will rotate at approximately 1500 rpm at this maximum power and 2400 rpm at 1KW. The vane motor will rotate at double this speed.

One of main advantages of air motors is their ability to stall under full load. When this happens there is little power loss in heat dissipation compared with that which would occur in hydraulic and electrical drives. However, the full torque would still be acting on the drive shaft which could cause extensive damage to plant. Another advantage of pneumatic drives are that they will operate satisfactorily in hazardous areas and where there is a large amount of water.

The disadvantages of air motors are that a sudden removal of the load can cause serious damage to the motor through overspeeding: as air is polytropic it tends to follow the Gas Laws when the load is removed. They will also discharge the lubricating oil as a mist into the surrounding atmosphere.

2.3 Hydraulic Drives

The power rating of a hydraulic drive can exceed 700KW at a speed of 1500rpm and this at a torque of 4653 Nm. Hydraulic systems provide numerous methods of combining speed and torque for a given reqirement.

2.3.1 Hydraulic drives using axial piston motors.

Axial piston motors are used mainly for relatively high speed applications. A small axial piston motor will rotate at 6000 rpm and produce 24 Kw of power. The flange size for such a motor will be less than 100 mm and the overall length less than 240 mm.

An area of application for axial piston motors is when a large concentration of power is required in a confined space. A 100 Kw electrical motor with an envelope size, in millimeters of 1315 x 600 x 890 cannot be fitted into a space of 400 x 400 x 210mm, whereas a hydraulic motor would fit comfortably. In fact it is possible to provide more than 750Kw of power on a space 400 x 400 x 900 mm. So a feature of fluid power drives is that they are compact. With hydraulic drives, the desired output speed can be achieved just by selection of the right components.

2.3.2 Hydraulic drives using radial piston motors

Radial piston motors are usually chosen for applications where large starting torques are required. It is not uncommon to have torque ratings of over 20,000 Nm. The radial piston motor is considered for slow speed high torque applications. The maximum continuous speed of a motor producing 20,000 Nm would be 100 rpm.

This type of motor would produce about 165 Kw at these ratings and take up 600mm diameter by 560mm of length. The swept volume of such a motor will be 6.8 litres per revolution.

So, small volume pumps can provide drives of high torque and slow speeds simply by choosing two stock items: a standard pump and a standard motor.

A variable displacement pump of 45 cc/rev. could produce a drive speed of between 1 and 5 rpm at a torque of 20,00 Nm.

So for standard drives, where one has to consider environment, space and speed, the first choice must always be fluid power. Additionally, if the application should have to have a stall facility without causing damage, then hydraulics and pneumatics should automatically be considered.

3. EFFICIENCIES OF THE VARIOUS DRIVE SYSTEMS

With all mechanical machines power losses are experienced due to friction, heat and restrictions etc. Each of the different types of drive will produce different efficiency levels when operated over a given torque/speed range.

3.1 Electrical drive losses

It is normal to find an AC. electric motor operating at approximately 85-90% efficiency at its rated speed. However, when one changes the speed the efficiency can reduce to 77%.

For instance: an electric motor may only give 25% of its rated torque when rotated at 100% above its rated speed. This means that there is a loss of power equal to 50%. Similarly at slow speeds the maximum torque can be 75% of its rated capacity, reducing the efficiency of this type of drive dramatically.

3.2 Pneumatic drive losses

It is always necessary to consider the speed range and torque requirements of air motors. These drives will range in output between 88-93% over their normal operating speed range. Unlike Electrical drives they do not have such an excessive power loss due to fall-off of torque or speed. Consideration must be given to the correct sizing of the valve, otherwise pressure drops across the valve will affect the difference between input and output power.

With pneumatic drives it is necessary to consider the power losses between the compressor and the valve. These losses are usually significant.

3.3 Hydraulic drive losses

When a hydraulic motor is used in a hydrostatic drive system, the operating efficiency of both the pump and hydraulic motor must be taken into consideration. It is usual to find that a pump will operate at approximately 92% efficiency and a motor between 88-94% giving an overall loss of power of 18%. Cooling and filtering will be needed which could increase this loss to 25% making a typical transmission approximately 75% efficient.

Hydraulic drives can be controlled by valves which will improve the direct losses in that the system will operate at the higher figure of the hydraulic motor, but pressure losses through the valve will decrease this to probably 80%.

© IMechE 1985 C277/85

Another popular way of control for a motor in hydraulic systems is to provide a combination of pressure and flow compensation to the pump which limits the output flow to the desired setting. This method of control is possibly one of the most efficient.

4 BENEFITS

The benefit of electrical drives is that they will use standard AC motors. They are clean and do not foul the atmosphere. They are compact in size and can easily be incorporated into computer controlled systems.

Pneumatic drives are somewhat more bulky in that they require air preparation and control equipment to enable them to be properly operated. Their main advantage over any other type of drive system is their ability to work in hostile environments. It would be sensible to use air motors in hazardous areas where an explosion could occur if electric motors were used. Similarly they will operate efficiently submerged in water, even at depths used in oil exploration.

Hydraulic drives can be used in similar situations to air motors provided the power pack can be adequately protected from hazardous gases.

Hydraulic motors have the ability to transfer large amounts of power in very compact sizes. Where space limits the other type of drive system, the hydraulic motor will always provide the answer. For example, a motor as big as a man's fist is capable of transmitting over 50Kw of power. The limitation usually is the space needed for the interconnecting pipe work.

5 COMPARATIVE TORQUE AND SPEED RATIOS

A comparison of the three main drive systems is shown (fig 3), with all of the values projected on a logarithmic scale. The greatest torque output is produced by the hydraulic motor. For high speed applications there is the axial piston motor.

The limitations of the electric motor drive can be seen in that it is limited to a band between 50-3000 rpm.

Although it is seen that the hydraulic motor starts at torque ratings which are greater than the eltecric motor, reducing the maximum pressure to the hydraulic system will lower the output torque.

The air motor fits into a similar envelope size to the electric motor. However, to see the full advantage of pneumatic drives they must also be compared with one another for price.

6 COST COMPARISONS

Figure 4 compares the costs of each of the drive systems. It is immediately seen that the pneumatic drive costs less than the other two systems. For a machine which requires a drive that is less than 20 Kw and will be used in a workshop or operating environment which already has air available then pneumatic drives would be the obvious choice.

An electrical drive for the same power would cost £3000 for the controller alone. For a system to control a 45 Kw electric motor the cost would be approximately £8000.

Hydraulic systems tend to be more expensive at the lower power ratings because they require oil reservoirs, filtration equipment and safety valves. However,if these items are readily available then the cost comparison improves. For £8000 it would be possible to provide a hydraulic drive capable of producing over 200 Kw of power.

7 COMPUTER CONTROL AND DRIVE SYSTEMS

Electrical engineers have quickly grasped the principles of computer control for operating electrical drives at variable speeds. In fact the computer has made it possible to produce an efficient electrical drive system.

Unfortunately, the fluid power engineer has been slower to see the benefits of computer control for either hydraulic or pneumatic drive systems. Only one manufacturer has developed a computer control system and this is not commercially available. The hydraulic equipment manufacturers have not as yet produced a computer control for hydraulic motors and so have experienced the loss of a whole market which they previously had to electrical drives. Admittedly this is using DC motor drives. Nevertheless, it is now possible to purchase electric cylinders.

The lack of expertise in electronics and computing by hydraulic equipment manufacturers has meant that hydraulic drives have not kept pace with modern technology. Unless investment is made in training staff, joint ventures with Universities and Technical Colleges, then the rapid changes being made in electronics will never be realized.

Universities and Technical Colleges must also become much more proficient in the latest developments in hydraulic equipment and teach the principles of fluid power design in a practical way.

8 CONCLUSION

It can be said that there is no hard and fast rule relating to the choice of a drive system for a given application, but certain factors can make the decision inevitable.

If it is cost that is the deciding factor then pneumatic drives must be our first consideration. If it is high power then hydraulic drives are the only choice. For integration into a computer system there is again no choice because only electrical drives currently have equipment available.

Unless heavy capital investment is made the fluid power industry will contract even further. Modern technology is here to stay and it must be incorporated into fluid power equipment so that it is available to the design engineer to produce better machinery. This is the only way to win exports, the only way to beat overseas suppliers and to restore confidence in products which are made in this country.

References:

The Total Control Package: Fenner International Ltd.

Fluid Power with Microprocessor Control: an Introduction. Reed/Larman. Prentice/Hall International. July 1985

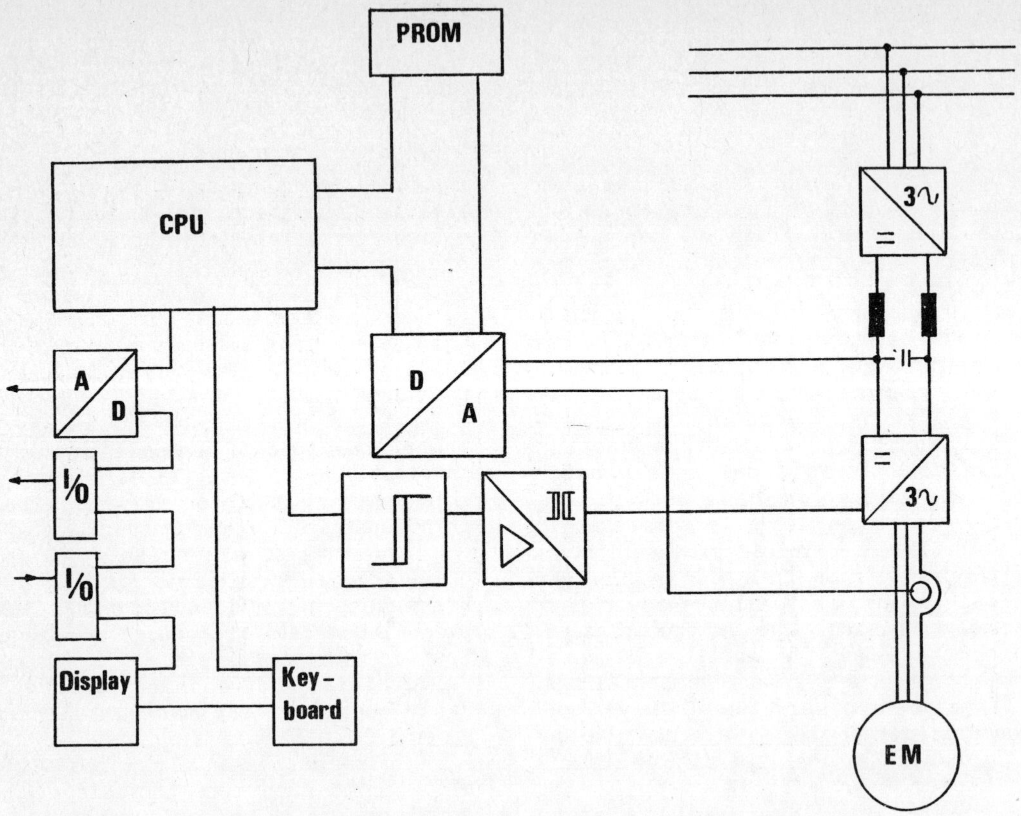

Fig 1 Block diagram showing the microcomputer control system

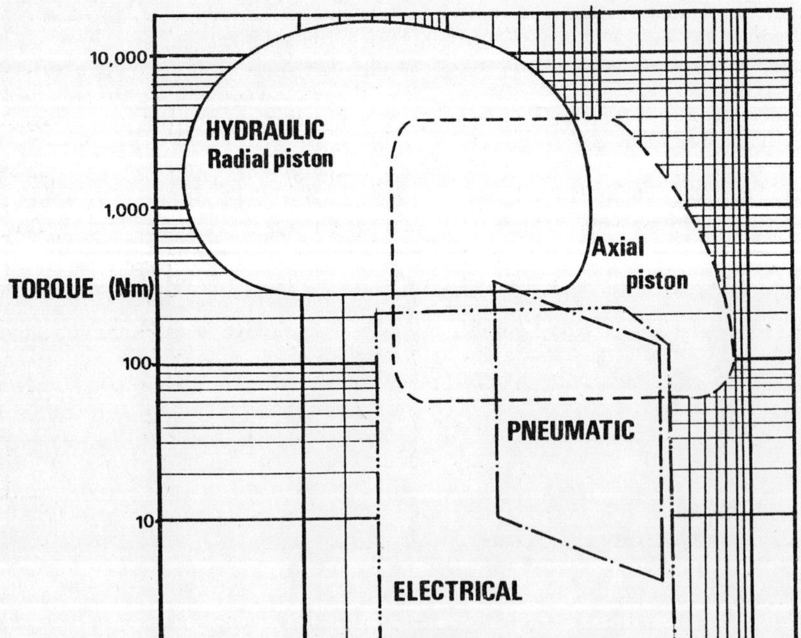

Fig 2 Comparison of torque and speed

© IMechE 1985 C277/85

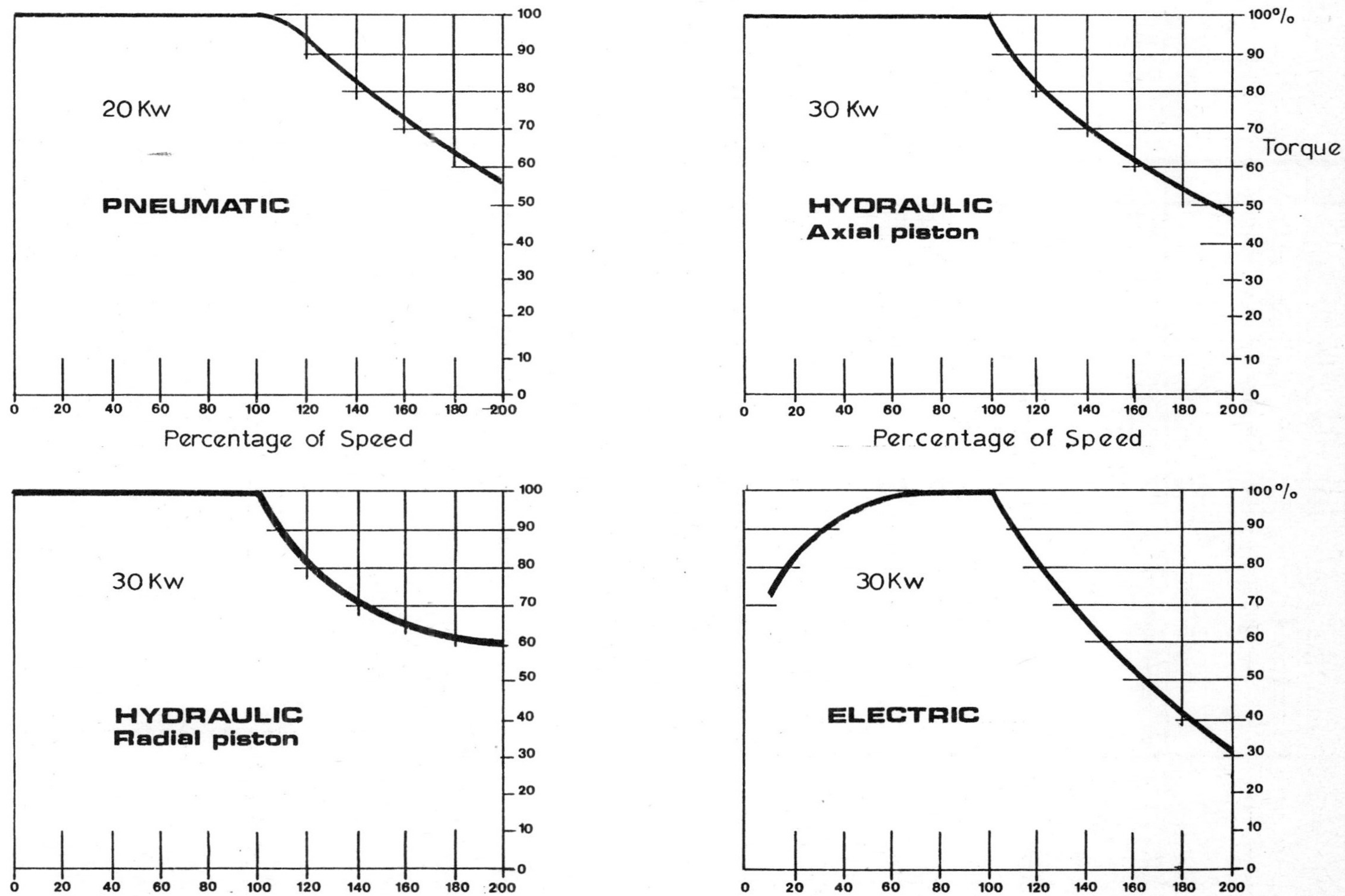

Fig 3 Comparison of torcue and speed – the three main drive systems

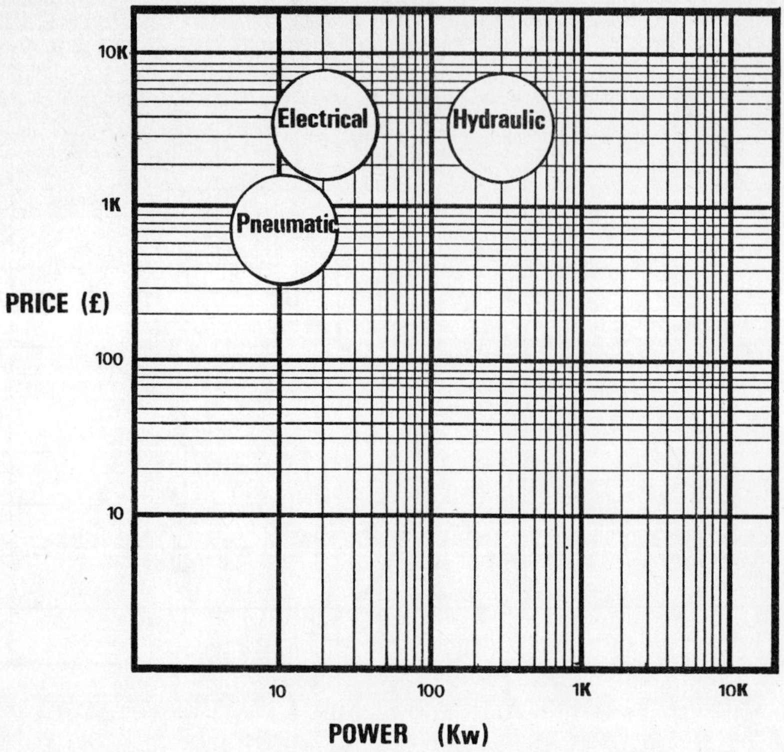

Fig 4 Comparison of unit cost with power

© IMechE 1985 C277/85